CHROMATOGRAPHIA

An International Journal for Rapid Communication in Chromatography, Electrophoresis, and Associated Techniques

Abstracted in Anal. Abstr., ASCA. Biodet. Abstr., Biol. Abstr., Cadscan, Chem. Abstr., Chem. Cit. Ind., C.I.S. Abstr., Current Contents, Deep Sea Res. & Oceanogr. Abstr., Diary Sci. Abstr., Excep. Med., Food Sci. & Techn. Abstr., GeoRef., INIS Atormind. Ind. Sci. Rev., Ind. Vet., Lead Abstr., Mass Spectr. Bull., Nat. Sci. Cit. Ind., Rev. Med. & Vet. Mycol., Sci. Cit. Ind., Sel. Water Res. Abstr., Sugar Ind. Abstr., Vet. Bull., VITIS, Weed Abstr., W.R.C. Inf., Zine Scan

Volume 53, Supplement 2001

Editorial Office

M. Schaub, Manager

Vieweg Publishing
P.O. Box 1546
65173 Wiesbaden, Germany

Tel. +49 (0)611 7878 380, 381
Fax +49 (0)611 7878 439

H. Weinheimer, Publisher

CHROMATOGRAPHIA

An International Journal for Rapid Communication in Chromatography, Electrophoresis, and Associated Techniques

Contents Supplement Vol. 53, 2001

Chromatographia CE Series
Series Editor: Kevin D. Altria, Analytical Evaluation
Group, Glaxo R&D, Ware, Herts SG 12 ODP, UK

**CE in Biotechnology: Practical Applications for Protein
and Peptide Analyses**
Edited by Anthony B. Chen,
Genentech Inc. South San Francisco, CA, USA

Wassim Nashabeh
Genentech Inc. South San Francisco, CA, USA

Timothy Wehr
Bay Bioanalytical Laboratory, Richmond, CA, USA

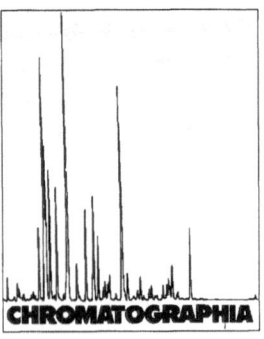

Vol. 53, Supplement 2001

Chromatographia was founded by R. E. Kaiser in 1968.

Publisher

Chromatographia is published by Friedr. Vieweg & Sohn Verlagsgesellschaft mbH, P.O. Box 1546, D-65173 Wiesbaden, Federal Republic of Germany, Tel. +49(0)611 7878 380(–381); Telefax +49(0)611 7878 439

Editorial office e-mail: CHROMATOGRAPHIA@bertelsmann.de; webpage: http:\\www.chromatographia.de

For more information regarding Vieweg's program for books and journals see our homepage: http://www.vieweg.de

Advertising Representatives

Inquiries concerning advertising should be addressed to the publisher's address above; Tel. +49(0)611 7878 153, Fax –430.

Inquiries in USA: Trade Media International, 424 Madison Avenue, New York, NY 10017, USA; Tel. (212)421–1229.

Inquiries in the UK: Elsevier Science Ltd., The Boulevard, Langford Lane, Kidlington, Oxford, OX5 1GB, UK.

Distributors

Friedr. Vieweg & Sohn, P.O. Box 1546, D-65173 Wiesbaden, Germany; Tel. +49(0)611 7878 324; Telefax +49(0)611 7878 423.

Elsevier Science Ltd., The Boulevard, Langford Lane, Kidlington, Oxford, OX5 1GB, UK.

Distributions in the USA

Chromatographia (USPS No. 374 810) is distributed by German Language Publications, Inc., 153 South Dean Street, Englewood, NJ 07631. Second class postage is paid at Englewood, NJ 07631.

Postmaster: send address changes to Chromatographia, German Language Publications, Inc., 153 South Dean Street, Englewood, NJ 07631.

Subscriptions

Chromatographia is published monthly. Up to three volumes may be published per year.

Vols. 53, 54, Supplements (2001) (approx. 2440 pp.)	DM 2.364,–	US $ 1,243.00	öS 17.257,–	sFr 2.103,–	€ 1.209,–
Single copy	DM 224,–	US $ 118.00	öS 1.635,–	sFr 199,–	€ 115,–

For individual subscribers who will certify that Chromatographia is for their personal use only (to be ordered directly from the publisher):

Vols. 53, 54, Supplements (2001)	DM 1.344,–	US $ 706.00	öS 9.811,–	sFr 1.196,–	€ 687,–

All prices include postage. Subscriptions are renewed automatically for one year unless notice to terminate the subscription is given three months before the end of the current year.

Submission of Papers

One original and two copies of manuscripts should be sent to Chromatographia, Editorial Office at the same address as the publisher, above. For papers intended for review-type articles, an outline of the proposed article should first be forwarded to the Editorial Office Manager for preliminary discussion, prior to preparation. For general information on the rules concerning style and format of manuscripts please refer to "Instructions to Authors" in every issue.

Softcover reprint of the hardcover 1st edition 2001

ISBN-13: 978-3-322-83023-4   e-ISBN-13: 978-3-322-83021-0

DOI: 10.1007/978-3-322-83021-0

Chromatogram on front page: Gas chromatographic separation of gasoline hydro carbons (selected section of chromatogram) with glass capillary column.

Preface

The continuing development of capillary electrophoresis (CE) is evident by the increasing number of journal articles, text books and symposia devoted to CE. This monograph is part of a continuing series covering individual areas of CE. A list of topics already published in this series includes: Analysis of nucleic acids by capillary electrophoresis (edited by Christoph Heller), Analysis of pharmaceuticals by capillary electrophoresis (written by Kevin Altria), Analysis of carbohydrates by capillary electrophoresis (written by Aran Paulus and Antje Klockow-Beck), Capillary Electrophoresis: instrumentation and operation (written by Wim Kok).

The Symposium 'CE in Biotechnology: Practical Applications for Protein Analysis' was held in San Francisco, California, USA on August 18, 1999. The symposium highlighted practical applications, including methods which had obtained regulatory approval for the routine analysis of marketed recombinant protein pharmaceuticals. This meeting demonstrated that CE has found a niche in the routine analysis of proteins and peptides in an industrial setting. The editors of this monograph, along with the series editor (Kevin Altria) were of the opinion that the proceedings of that meeting could provide the basis for theoretical and practical issues covering the analysis of proteins and peptides by CE. This supplement therefore consists of some of the practical applications that were presented at the symposium and several review chapters, including 'CE in the development of recombinant protein biopharmaceuticals', 'Selection of buffers in CZE: application to peptide and protein analysis' and 'Recent advances in cIEF'.

The editors would like to acknowledge Dr Barry L. Karger, who encouraged us to hold the CE Symposium (in his words, the time was right), the California Separation Science Society for sponsoring the symposium, and Michael Kunitani and SungAe S. Park who were also on the organizing committee. We would like to thank the symposium participants for their efforts in preparing manuscripts. The editors greatly appreciate the efforts of Amy Greene-Dittz who reviewed all the figures for the manuscripts, scanned in images when electronic copies were not available and modified some images to increase their legibility. We would also like to thank Angelika Schulz for her patience and help in assembling this monograph.

Anthony B. Chen
Wassim Nashabeh   August 2000
Timothy Wehr

CHROMATOGRAPHIA

CE Series

Edited by Kevin D. Altria, Glaxo Wellcome R&D, UK

There are currently a number of general textbooks covering Capillary Electrophoresis where information on a range of applications and techniques can be found. Readers who are interested in a specific area of CE struggle to find truly comprehensive treatments of their areas of interest. The CHROMATOGRAPHIA CE series has been established to allow comprehensive books to be produced covering individual topics. The books are written by well known authors in their specialist application areas and cover CE topics such as DNA analysis, analysis of pharmaceuticals, chiral separations, MECC, carbohydrate analysis, biomedical applications and troubleshooting in CE.

- **Volume 1:** C. Heller (Ed.), Analysis of Nucleic Acids by Capillary Electrophoresis

- **Volume 2:** K. D. Altria, Analysis of Pharmaceuticals by Capillary Electrophoresis

- **Volume 3:** A. Paulus / A. Klockow-Beck, Analysis of Carbohydrates by Capillary Electrophoresis

- **Volume 4:** W. Kok, Capillary Electrophoresis: Instrumentation and Operation

- **Volume 5:** A. B. Chen / W. Nashabeh / T. Wehr (Eds), CE in Biotechnology: Practical Applications for Protein and Peptide Analyses

Capillary Electrophoresis in the Development of Recombinant Protein Biopharmaceuticals

A. B. Chen[1] / E. Canova-Davis[2]

Departments of Quality Control[1] and Analytical Chemistry[2], Genentech Inc. One DNA Way, South San Francisco, CA 94080, USA

1.1 Summary

The understanding of the physicochemical properties of proteins in solution is required to understand the separations taking place in the capillary. For example, during isoelectric focusing, it is important to ensure the lack of protein precipitation during focusing and mobilization otherwise artifacts occur and the true resolution of charge variants is not achieved. Similarly, buffer conditions, including additives, composition of buffer, and pH can influence the resolution of product variants. With diligent efforts one can achieve desired separations. As knowledge of buffer conditions and use of these techniques increase, the methods will be improved.

MECC can be useful for the characterization of recombinant proteins and monoclonal antibodies. However, the relatively long analyte migration times may be a deterrent for routine use in quality control situations. Pharmaceutical companies will employ new techniques which provide advantages over current technology. CE has advantages in certain applications. Applications which use slab gel techniques, IEF and SDS-PAGE can be replaced by the CE methods for formulation development, characterization of charge isoforms, and for purity determinations. CZE and cIEF can be used for rapid identity tests which do not require peptide mapping. CZE has a tremendous advantage in analyzing basic proteins, and CZE and CE-SDS can be used to quantify product in the presence of interfering excipients. It is the responsibility of the analytical chemist to determine applications which are appropriate for CE. It is apparent that these are currently being defined for the biotechnology industry, and the usage of CE should continue to increase.

1.2 Introduction

Since the cloning and expression of insulin, the biotechnology industry has experienced a boom in the number of products that have been developed, tested in clinical trials, and approved for unmet medical needs for many indications. The development of proteins as pharmaceuticals is somewhat different than the development of traditional small molecule pharmaceuticals. These differences are mainly due to the size and complexity of proteins. With the experience gained after the approval of recombinant proteins such as insulin, growth hormone (rhGH), recombinant human erythropoietin (rhEPO), recombinant tissue plasminogen activator (rt-PA), and a number of monoclonal antibodies, the required characterization and process controls for the approval of a protein product are becoming more established. After the protein has been cloned and expressed in eukaryotic or prokaryotic cells, there are usually several rounds of purification, product characterization, evaluation of the production process, and preclinical and toxicology studies leading to the filing of an IND to begin clinical trials. Assays are required for quantitation of the protein during various stages of production including the crude cell paste or cell culture fluid to determine the yield of the production process. During production, various post-translational modifications can occur such as deamidation, oxidation, isoaspartate isomerization, and proteolytic cleavages. Hence, it may be necessary to analyze for size, charged, or hydrophobic variants, especially if these variants have different biological potencies than the desired product. According to the ICH guideline, Q6B 'Guidance on Specifications: Test Procedures and Acceptance Criteria for Biotechnological/ Biological Products' [1], a product variant which does not have the same safety and efficacy profile as the desired product is termed a product-related impurity. For example, isomerization of a heavy chain aspartate residue of a therapeutic monoclonal antibody to isoaspartate resulted in a variant with low potency [2]. The manufacturer may choose to remove these impurities during the purification process or monitor their concentration during lot release.

Product characterization is required throughout the development of a recombinant product. In early preclinical stages, it is necessary to confirm if the product manufactured is the desired product prior to beginning pharmacokinetic and toxicological studies. In the case of glycoproteins, there could be heterogeneity due to varying degrees of glycosylation and occupancy of glycosylation sites. These variations could affect the intrinsic biological activity of the protein or the pharmacokinetic profiles of these proteins. For exam-

0009-5893/00/02   7-11 $ 03.00/0

ple, it has been shown that glycosylation is necessary for the antibody-dependent cell-mediated cytotoxic and complement-dependent cytotoxic activities of monoclonal antibodies [3]. Carbohydrate moieties play a role in the clearance of rt-PA [4] and biological activity; e. g. type II rt-PA with an unoccupied glycosylation site displays more in-vitro activity than type I rt-PA which has all three glycosylation sites occupied [5]. In these instances, it may be necessary to monitor glycosylation to demonstrate consistent manufacture of a quality product. When the results of clinical trials are promising, more extensive characterization is usually performed for regulatory filings to demonstrate that the manufacturing process is well understood in order to avoid surprises when full production occurs. Some of the methods that are used in characterization are validated for final product lot release testing. Analytical characterization is also required to support methods for use in the quality control system. Formulation studies are performed to optimize the excipients and conditions necessary for the stability of the recombinant protein over a proposed shelf-life. Formulation studies are also performed to determine the modes of degradation of the protein molecule. Thus, formulation studies normally include accelerated stability studies of the protein molecule under various conditions. Following successful clinical trials and the finalizing of the manufacturing process, validation of the methods to be used in the quality control of the product is initiated. The control system for lot release is determined based upon product characterization and manufacturing history. Tests for identity, strength, purity, and potency are among those required for final product lot release. The primary identity test used requires peptide mapping to confirm the amino acid sequence of the protein. Identity tests are also used to confirm the product's identity, for example during packaging, or transfer of product to collaborators and contract labs. Stability studies are required on bulk and final product to confirm the shelf life of the recombinant product.

The traditional chromatographic methods such as reversed-phase high performance liquid chromatography (RP-HPLC), size-exclusion chromatography (SEC), ion-exchange chromatography (both cation and anion), hydrophobic interaction chromatography, and gel electrophoretic methods such as sodium dodecylsulfate polyacrylamide gel electrophoresis (SDS-PAGE) and isoelectric focusing (IEF) methods have been used extensively for the analysis of recombinant proteins. CE analyses add to the repertoire of available analytical methods and will be discussed as they support the requirements for product development.

1.3 Capillary Electrophoresis Reviews

Several excellent reviews have been written in the use of capillary electrophoresis for the analysis of proteins including biotechnology-derived proteins. These reviews have provided information based upon the individual capillary electrophoresis (CE) techniques such as capillary zone electrophoresis (CZE, [6–8]), capillary isoelectric focusing (cIEF, [9, 10]), capillary electrophoresis using sodium dodecyl sulfate (CE-SDS, [11]) and micellar electrokinetic chromatography (MEKC, also called micellar electrokinetic capillary chromatography [MECC], [12]). Reviews have also been written for the analysis of recombinant proteins in general [13–15], antibodies [16], and glycoproteins [17, 18].

1.3.1 CZE

The simplest mode of capillary electrophoresis is that of free solution capillary electrophoresis or CZE. In this mode, the intrinsic properties of the molecule, its charge and size are used to characterize its electrophoretic behavior, determine its purity, and as a means of quantification. In Dolnik's review [6] of the CZE of proteins, he addressed modeling the migration behavior of proteins, sample pretreatments such as preconcentration and derivatization to increase sensitivity, methods to reduce interactions with the capillary wall, and ways to improve the selectivity of the method. He also briefly reviewed CE-SDS using cross-linked and non-crosslinked gels. When more sensitive methods than UV detection are required, laser-induced fluorescence (LIF) or mass spectrometry methods have been used. The detection limit for bovine serum albumin (BSA) decreases from $0.5\ \mu g\ mL^{-1}$ to $25\ ng\ mL^{-1}$ when UV absorption is replaced with argon-laser-induced fluorescence [19]. In his subsequent review, Dolnik [7] updated the information on these same topics. Preconcentration can be performed by three on-line methods. In the first, sample self-stacking was used to determine trace concentrations of recombinant interleukins [20]. The second procedure utilizes solid-phase extraction on a cartridge containing Spherisorb C18, with the concentrated peptides or proteins being released with acetonitrile. The latter method was used to analyze a tryptic digest of bovine serum albumin. The third method for preconcentration used a semipermeable hollow fiber between the sample vial and the capillary inlet. An injection current is applied which concentrates the proteins into the hollow fiber. Subsequently, an electric field is applied through the hollow fiber to begin the electrophoretic separation. This method can lower the detection limit by a factor of 1000 [21]. Because the walls of fused silica capillaries carry negative charges due to ionized silanol groups, some proteins may adsorb to the walls during electrophoresis. Either dynamic (buffer additives) or static (permanently coated) wall coatings are used to reduce adsorption to capillary walls. For improving the selectivity of a CZE method, optimizing the composition of the background electrolyte (BGE) is usually a more effective procedure with adjusting the pH of the BGE as the first choice. A considerable amount of research has been done in describing buffer additives for CZE which minimize protein-capillary interactions, improve selectivity and resolution, and control electroosmotic flow (EOF). Corradini [22] has summarized these additives which include neutral polymers, ionic salts and zwitterions, amine modifiers, surfactants, and ion-pairing agents used in protein CZE methods. In some cases cylcodextrins have been successfully used to resolve protein mixtures. Thus, carboxymethylated β-cyclodextrin was used to improve the separation of a model mixture of proteins (α-chymotrypsinogen A, cytochrome c, lysozyme, and ribonuclease A) in the cationic mode at pH 2.5.

A fundamental requirement in the analysis of a recombinant protein pharmaceutical is the need to demonstrate identity with respect to its primary amino acid sequence by peptide mapping. The method may also be used to show, by comparison with an appropriate reference material that changes in the primary amino acid sequence have not occurred, confirming product consistency and/or genetic stability [23]. The method requires considerable expertise for performing peptide analysis in a quality control environment [23]. At present, only peptide maps generated by HPLC have been reported to be used for regulatory filings. Considerable progress has been made in the analysis of peptides by CZE, and this is one of the most promising areas for routine analysis in quality control. Janini and Issaq in this monograph [24] provide the background for selection of a buffer to perform CZE analysis of proteins and peptides. This chapter provides a detailed description of buffer basics including temperature and Joule's heat considerations, buffer type and concentration, and isoelectric buffers. The authors noted that in CZE some buffer-capillary wall interactions can enable high resolution of peptides and buffer-analyte interactions can result in different electrophoretic profiles for the same sample depending upon the buffer system used. The buffers that have been used to perform CZE of peptides and proteins were presented, and buffer additives that have been used to enhance resolution of CZE methods, control electroosmotic flow, and reduce analyte adsorption to the capillary wall were described. MEKC for peptide separation is also discussed. Janini et al. [25] measured the electrophoretic mobility of 58 peptides ranging in size from 2 to 39 amino acids and varying in charge from 0.63 to 7.82. Measurements were obtained with a polyacrylamide-coated capillary using a 50 mM phosphate buffer, pH 2.5, at 22 °C. The results indicated that the Offord model (26) which correlates electrophoretic mobility with the charge-to-size parameter $q/M^{2/3}$ provided a good fit of the experimental data. Endoproteinase Lys-C peptide maps of melittin (26 amino acids) and horse myoglobin were compared to theoretical simulations of mobility. The melittin comparison was excellent with respect to line position and relative area percent of the peptides. The agreement of the theoretical electropherogram of myoglobin although not perfect agreed reasonably well with the experimental data. The chapter by Chen and Fausnaugh-Pollitt [27] in this monograph compares the separation of 33 polycationic protegrin analogs by RPHPLC and CZE. These analogs (14 to 18 amino acids in length) consist of single amino-acid substitutions, truncations, cyclic, and D-amino acid analogs. The separation by CZE used a bare fused-silica capillary and 100 mM sodium phosphate buffer, pH 2.6, and was achieved due to charge/mass differences of the pep-

tides. The peptide analog mobility fit the Offord [26] equation. RPHPLC exhibited better separation of single D-amino acid substitutions, while CZE was better for truncated analogs. Thus, RPHPLC and CZE provided orthogonal methods for analysis of these peptides. Righetti et al. [8] describe the CZE of peptides and proteins in acidic, isoelectric buffers. These buffers enable the use of high voltage gradients (up to 1000 V cm^{-1}) in large bore capillaries which produce separations of a few minutes with high resolution. Thus, isoelectric aspartic acid (Asp) was used to generate a peptide map of trypsin digested β-casein. The pH of this isoelectric buffer ranges from 3.36 to 2.80 as its concentration is increased from 5 mM to 50 mM at 25 °C. Isoelectric Asp (p*I* 2.77 at 50 mM and 25 °C) containing 7 M Urea and 0.5% hydroxyethyl cellulose, could be used to analyze storage proteins in cereals such as gliadins in wheat and zeins in maize. The hydroxyethyl cellulose probably provided a dynamic coating to prevent the proteins from adsorbing to the capillary wall. Verzola et al. [28] describe a method for quantifying protein adsorption to bare fused silica and for determining the effect of amino quenchers added to the BGE. The method consists of flushing a fluorescently labeled protein into the capillary, equilibrated in Tris-acetate buffer, pH 5.0, until saturation of binding sites occurred. SDS micelles are then used to electrophoretically desorb the bound protein, which is quantified fluorometrically. Using fluorescein isothiocyante (FITC)-myoglobin to saturate protein-binding sites, Verzola et al. [28] determined that oligamines (spermidine, spermine, tetra-ethylenepentamine) were more effective than monoamines (triethylamine, triethanolamine, ethylamine) and diamines (putrescine, cadaverine, hexamethonium bromide) in quenching interactions with the silica wall. Using this method, these investigators also determined that electrophoretic desorption with SDS micelles (60 mM) was 100 percent effective for protein desorption, which was not the case for the commonly used washing procedures of 1 M NaOH and 1 M HCl.

1.3.2 cIEF

Wehr et al. [9] presented an excellent review on the capillary isoelectric focusing (cIEF) of proteins. They described the influence of sample preparation including

sample salt levels and protein concentration, and the importance of ampholyte composition depending upon the desired separation range. Coated capillaries are necessary to reduce electroosmotic flow which would result in band broadening. Once the proteins are focused, they must be mobilized for detection. Several examples are provided for chemical, hydrodynamic, and electroosmotic flow mobilization used to monitor the charged isoforms. Protein precipitation is described as the major source of difficulty in cIEF. Reducing the focusing time, lowering the protein concentration, and the addition of protein-solubilizing agents such as organic modifiers and surfactants are suggested for reducing protein precipitation. In that review the authors presented the cIEF of a mixture of standard proteins: rt-PA types I and II, iron-complexed transferrin, and hemoglobin mutants. Wehr et al. [9] concluded that cIEF compared favorably with conventional gel IEF in terms of selectivity and analysis time with the added benefit of complete automation. In a subsequent review, Rodriguez-Diaz et al. [10] included a discussion on the ampholyte composition, additives which influence the cIEF of proteins, the use of internal standards, and presented examples of cIEF of immunoglobulins and rt-PA wherein many aspects of the methodology were reviewed in detail.

Liu et al. [29] described cIEF as a tool for examination of peptides and proteins of pharmaceutical interest. These authors provided a detailed review of flat-bed IEF and cIEF, including theory and fundamentals and a comparison of these methods. They discussed the chemical nature of carrier ampholytes, the formation and stability of pH gradients, and the rationale for their focusing properties and described resolution and peak capacity in cIEF. Liu et al. [29] provided examples of the cIEF of hemoglobin, transferrin, and monoclonal antibodies. Conti et al. [30] addressed the issue of protein solubility in cIEF, describing a family of protein solubilizers (non-detergent sulfobetaines in concentrations up to 1 M) for preventing protein precipitation and aggregation at the p*I* value and at low ionic strength. The proteins that were used in these studies were L-aspartate oxidase, thermamylase (an α-amylase from Bacillus licheniformis), and alcalase (an alkaline protease of the subtilisin family). Conti et al. [30] suggested that common zwitterions such as taurine and a few of Good's buffers such as Bicine and CAPS were useful in acidic pH gradients up to pH 8.0. Addition of up to 20% of sugars, such as saccharose, sorbitol, and sorbose, improved protein solubility near the p*I* especially if the sugars were mixed with 0.2 M taurine. Tran et al. [31] described a one-step cIEF method for resolving the glycoforms of recombinant human immunodeficiency virus (HIV) envelope glycoprotein, rgp 160sMN/LAI. Separation was obtained with a polyvinyl-alcohol (PVA) coated capillary using a mixture of narrow and wide-range pH ampholytes. They determined that a mixture of a sugar, saccharose and a zwitterion, 3(cyclohexylamino)-1-propanesulfonic acid was the most efficient additive to avoid protein precipitation. The cIEF method was able to differentiate the rgp 160 glycoform patterns of two sub-populations of HIV-1. cIEF is also reviewed in this monograph by Wehr et al. [32] who provide up-to-date information on ampholyte selection, mobilization techniques, the use of internal standards, methods for desalting and preventing protein precipitation, imaged cIEF, and details of cIEF coupled to mass spectrometry. These authors also review in detail the cIEF of monoclonal antibodies, hemoglobin and hemoglobin variants, and erythropoietin.

1.3.3 CE-SDS

Sodium dodecylsulfate polyacrylamide gel electrophoresis (SDS-PAGE) has been extensively used for the analysis of recombinant proteins during formulation development, purification of the product, and as a general purity test for quality control. CE methods based on similar principles were reviewed by Guttman [11]. The CE technique has been called capillary sodium dodecylsulfate gel electrophoresis or capillary gel electrophoresis (CGE). The term, capillary electrophoresis SDS non-gel sieving (CE-SDS-NGS), has also been used when noncross-linked replaceable matrices were used as the sieving medium. In this chapter, CE using SDS for size separation of proteins will be referred to simply as CE-SDS. Guttman [11] described the theory of size separation using CE-SDS and evaluated the effects of operational variables such as field strength, temperature, capillary dimensions, and separation medium on the method. He reviewed the work performed using two types of gels that have been used, chemical and physical gels. An example of a chemical gel is cross-linked polyacrylamide. This type of gel is attached to the capillary wall. Size separation of proteins up to 35 kDa in molecular weight have been reported in chemical gels. An example of the powerful capabilities of this system is the separation of the two chains of insulin. The disadvantage of the system is that only clean samples can be injected otherwise the capillaries could clog and result in poor separation. Physical gels use non-cross-linked linear or slightly branched polymers as the sieving matrix and are replaced after each run. With the physical gel, pressure injection can also be used as a means of introducing the sample into the capillary. This option is not available with the chemical gels. The noncross-linked replaceable gel is becoming the preferred method for performing size-separation analyses with CE. These gels are commercially available. Guttman [11] compared SDS-PAGE and CE-SDS and concluded that similar separation results and precision can be obtained by either technique. With respect to molecular weigh determination, Guttman [11] presented a table comparing slab gel electrophoresis to CE-SDS on 65 individual proteins ranging in molecular weight from 14–206 kDa. These data clearly showed that glycoproteins behave anomolously, and if accurate molecular weights were required, the Ferguson method [33] described in Guttman's review must be used. However, for recombinant proteins, the CE-SDS method is used as a general purity test and for analysis of production samples in a similar manner to the use of SDS-PAGE. In this monograph, the chapter by Schenerman and Bowen [34] addresses the optimization, validation, and use of CE-SDS in quality control testing.

1.3.4 MEKC

In MEKC/MECC [12], surfactants such as SDS are added to the running electrolyte at concentrations above the critical micellar concentration. The electrophoretic migration of proteins is influenced by protein-micelle interactions that can give a charged micellar characteristic to a neutral protein. A highly significant benefit of the presence of surfactant is the elimination of protein-wall interactions. SDS may also induce protein denaturation, which could facilitate resolution of individual components. Separations of a crude E. coli fermentation broth of a recombinant protein were obtained using a 50 mM sodium borate buffer, pH 9.0, containing 0.3% SDS. The recombinant protein migrated after a large mass of cellular components which demonstrates the unusual selectivities that can be obtained with MEKC [35]. Another example is the use of MEKC to separate savinase (SAV, a serine protease) and SAV* forms which differ from SAV in a single residue wherein methionine is replaced by serine [36]. If investigation must be performed on the native protein, MEKC can be conducted under non-denaturing conditions using zwitterionic or non-ionic surfactants. The work done by James et al. [37] on interferon-gamma glycoforms using 400 mM borate and 100 mM SDS, pH 8.5, demonstrated that the glycoform migration times were inversely related to the amount of associated carbohydrate. The separation efficiency was a result of the synergistic action of both the reduced electroosmotic flow (EOF) and the higher surfactant concentration of the MEKC method as separate entities eliminating protein adsorption. James et al. [37] suggested that the analytes with the shortest migration times had the highest carbohydrate content and thus the largest glycan structures. As the concentrations of borate and SDS in the buffers were increased, there was an increase in resolution observed, and three main peak groups were resolved. Kats et al. [38] used MECC to separate four major isoforms of the BR96 antibody following heat treatment of the molecule in the presence of 25 mM SDS in 12 mM sodium borate buffer, pH 9.4. These analyses were correlated to changes observed by circular dichroism spectropolarimetry.

Recently, Miksik and Deyl [39] published their investigation on separation of proteins and peptides in acid buffers containing high concentrations of surfactants (SDS). SDS was used because it has a washing effect on the capillary wall. Secondly, even under acidic conditions SDS can bind to protein analytes which then bear considerable negative charge that enhances their separation. Apparently this system is suitable for proteins which are soluble in acidic buffers. Separation of four standard proteins, lactalbumin, cytochrome c, carbonic anhydrase, and urease was presented as well as the separation observed with a leucocyte lysate preparation.

1.4 Reviews of CE Used for the Analysis of Recombinant Proteins

The CE of recombinant proteins was reviewed in 1997 by Denton and Tate [13] and by Strege and Lagu [14]. From the aspects of product development, Denton and Tate's review covered the characterization of recombinant proteins using CE; for example, comparison of natural and recombinant proteins, separation of mutant and variant forms, analysis of glycoforms, and confirmation of protein identity as defined by peptide maps. Due to the high resolving power of CZE, certain protein modifications can be resolved in some cases by analyzing native proteins; for example, acetylation of insulin, deamidation/isomerization of rhGH, and oxidation/reduction of human serum albumin (HSA, [13]). CE (CZE, MECC, or cIEF) was especially useful for the analysis of protein glycoforms and the determination of the carbohydrate microheterogeneity of glycosylated proteins. Denton and Tate presented results showing the separation of mutants and variants which may be difficult to resolve by current analytical methodology. For example, four IGF-I variants with closely related mass-to-charge ratios could be separated due to selectivity obtained using a zwitterionic detergent as a hydrophobic selector [40]. A neutral-coated capillary was used to minimize EOF and analyte-wall interactions [40]. Twelve Staphylococcal nuclease variants obtained by site-directed mutagenesis at a single amino acid residue could be resolved using a capillary coated with an inert hydrophilic layer and oligoamino buffers between pH 2.8 and pH 9.5 [41]. Glycoforms of bovine pancreatic ribonuclease B with different numbers of mannose residues were resolved on a fused-silica capillary with a sodium phosphate-sodium tetraborate buffer, at pH 7.5, containing 50 mM SDS [42]. Watson and Yao were able to separate rhuEPO into at least six glycoforms using an uncoated fused-silica capillary with a tricine buffer at pH 6.2, containing 1,4-diaminobutane (DAB) to reduce EOF, and 7 M urea for resolution and peak shape [43]. DAB was also used as a buffer additive to resolve recombinant granulocyte colony stimulating factor (rGCSF, [44]) and factor VIIa glycoforms [45] and was therefore useful for resolving sialic acid-containing variants. Yim et al. [46] used a phosphate buffer system at pH 2.5, with no additives, to analyze nine glycoforms of recombinant human bone morphogenic protein 2 (rhBMP2). Yim et al. demonstrated that rhBMP-2 consists of 15 glycoforms, but only those with differing numbers of mannose residues could be separated. Glycoforms with the same number of mannose residues (ie. stereoisomers) could not be separated.

Peptide maps are a regulatory requirement to demonstrate the identity of the protein and are also used to demonstrate genetic consistency in recombinant DNA-derived proteins [23]. Peptide maps have also been used for determining changes in oxidation, deamidation, and the primary sequence of proteins by HPLC. CZE was used to generate a peptide map of recombinant growth hormone (rhGH, [47]). CZE displayed an advantage over HPLC in greater throughput of samples and a decrease in analysis time. Both HPLC and CZE separated nineteen hGH tryptic peptides but the order of elution was different. Thus, the complementary use of CZE and HPLC could provide a more complete peptide analysis by further resolving peptides and confirming single peaks. Similarly, sixty-five peptides of non-reduced HSA could be resolved by CZE [13]. A system for predicting peptide mobilities in CZE separations was developed which uses the pK_a of a peptide at a specific pH to calculate mobility at a given voltage [48]. This system is useful for determining peptide identity.

Strege and Lagu [14] summarized the analyses of biotechnology-derived proteins using cIEF, MEKC, CE of pharmaceutical dosage forms, CE/MS, peptide mapping, and CE-SDS. This review examined in considerable detail the characterization of rt-PA by cIEF. Optimization efforts, methods, and assay parameters were described by several investigators. rt-PA is a complex glycoprotein consisting of approximately 8–9 peaks on cIEF due to sialic acid heterogeneity from N-linked glycosylation sites. Deamidation of asparagine residues can also add to the charge heterogeneity. The impact of carrier ampholytes on the cIEF separation were described by Chen et al. [49], Kubach and Grimm [50], and Thorne et al. [51] in their analyses. Some sources of carrier ampholytes appeared to have background peaks and could be a source of nonprotein peaks. Urea appeared to be necessary for resolution of rt-PA glycoforms [50, 51]. Thorne et al. described an ELISA method for demonstrating good recovery of the protein from the capillary during CZE and cIEF of rt-PA. They showed that rt-PA could be characterized into its glycoform variants using CZE on a coated capillary and in an ε-aminocaproic acid buffer containing 0.01% Tween 80 buffer. The difference between major variants (type I and type II) of rt-PA could be distinguished by cIEF and by CE-SDS. Frenz et al. [52] demonstrated the complementary analysis of tryptic peptides of hGH by HPLC and CZE, and Rush et al. [53] used CZE for peptide mapping of recombinant human erythropoietin (rhEPO) suggesting that the CZE method could be validated for product lot release. Yowell et al. [54] achieved separation of recombinant GM-CSF from the formulation components using both the one-step cIEF method and CZE. Therefore, these methods could be used to determine concentration (strength) of the active component in the final dosage form. The validation of a one-step cIEF method for demonstating consistency of rt-PA glycoforms was attempted by Moorhouse et al. [55]. Protein recovery from the coated capillary was observed to be >90%, the limit of detection was determined, and the precision of the area percent shown to be acceptable. However, due to the variability of residual EOF in different lots of the coated capillary, resulting in highly variable migration patterns and migration times, these capillaries were deemed unacceptable [55]. Hunt el al. [56] reported on cIEF of a monoclonal antibody to be used as an identity test for quality control purposes. The cIEF profile was correlated to the five components observed on gel-IEF. As a general purity test, CE-SDS was suggested as a replacement for the more cumbersome SDS-PAGE gels. Hunt et al. (56) showed the limits of detection were approximately 0.5–1 ppm with UV detection. This was better than the sensitivity of Coomassie-stained gels but not quite as sensitive as silver-stained gels. The method was also stability indicating as low molecular weight peaks increased with storage as the main peak was decreased.

In 1999, Lagu [15] reviewed the applications of CE in biotechnology focusing on antisense oligonucleotides and rDNA-derived proteins. He discussed the analysis of recombinant proteins from cell lysates, presenting the work of McNerney et al. [57] analyzing rhGH in E coli extracts. Using a phosphate-deactivated capillary and a phosphate buffer system (250 mM phosphate buffer at pH 6.8 containing 1%

propylene glycol), McNerney et al. [57] were able to resolve several variants of rhGH, including deamidated and N-terminally truncated forms, from the desired product. Klyushnichenko et al. [58] extracted the enzyme NADP$^+$-dependent formate dehydrogenase from E. coli cell pellets with phosphate buffer and heated the supernatant with a solution containing SDS and dithiothreitol before analyzing the extracted enzyme by CE-SDS and SDS-PAGE. These methods were used to follow the production of enzyme as a function of time. MEKC, using an uncoated silica capillary and SDS/borate buffer, was used to monitor glycosylation patterns of interferon-γ in Chinese hamster ovary (CHO) cells, under various fermentation conditions. The three major groups of variants arising from the occupancy of two glycosylation sites could be resolved and quantified. The relative proportions of these variants were constant during culture. cIEF resolved the recombinant IFN-γ into at least 11 different glycoforms. During perfusion culture the acidic IFN proteins increased after 210 hours of culture, indicative of N-glycan sialylation [59].

The carbohydrate of rhEPO plays a vital role in biological activity; therefore, it is important to be able to characterize that portion of the molecule. rhEPO contains three N-glycosylation sites with a variable number of sialic acid residues. Cifuentes et al. [60] observed seven glycoform peaks by cIEF which corresponded to agarose gel IEF bands. cIEF of rhEPO was not a simple task as total desalting of the protein led to a poor resolution; hence, some residual excipient in 7 M urea was necessary for the optimum separation. Cifuentes et al. [60] provided a comprehensive review of the cIEF work performed to date on rhEPO. These authors were able to obtain the same separation using CZE with better precision in separating the isoforms. They stated that the CZE method did take a longer time to perform. The pI range of the isoforms was 3.78 to 4.69. In contrast to flat bed determinations, the CE methods were quantitative. HSA is included in the EPO formulation buffer, and the two proteins can be resolved by CZE using an amine-coated capillary and 200 mM phosphate buffer, pH 4.0, containing 1 mM nickel chloride [61]. Therefore, this method can be used to quantify rhEPO in final product. Another CZE method was reported wherein an uncoated capillary and 20 mM phosphate

buffer, pH 6.0, containing 5 mM putrescine was used to resolve rhEPO glycoforms [62]. Zhou et al. [63] analyzed rhEPO from three different manufacturers using CZE, HPLC, MALDI-TOF-MS, on-line CE-ESI-MS, and LC-ESI-MS. These analyses were performed on both the intact protein and tryptic digests. Using CZE, six protein-related peaks could be detected in the intact protein due to charge heterogeneity arising from the variable sialic acid content. Products from the same manufacturers showed little lot-to-lot variation in their glycoform patterns indicating good consistency of the manufacturing processes. Appreciable differences in shape and relative proportions of the main peaks were observed between manufacturers.

Farchaus [64] examined an antigen protein which is a major component of human anthrax vaccine by SDS-PAGE and CE-SDS. The purity analysis by CE-SDS allowed the estimation of contaminants in the purified protein. These impurities ranged from 1–3% and could not be quantitated by SDS-PAGE. Roddy et al. [65] developed a CZE method for purity determination of acidic fibroblast growth factor (FGF). This method included phosphate buffer at pH 2.5 and hydroxypropyl methyl cellulose. This was a particularly significant achievement since acidic FGF tends to precipitate.

Krull et al. [16] reviewed the existing literature on the CE methods for the identification and quantitation of antibodies, conjugates, and complexes. The authors described the analysis of antibodies using the major modes of CE and the major application of the respective modes, including cIEF, MEKC, and CE-SDS. They discussed the achievement of resolution by these modes of CE. Five major peaks were observed for a recombinant monoclonal antibody in IEF and cIEF by Silverman et al. [66]. Pritchett et al. [67] reported six peaks for an anti CEA MAb, and Hunt et al. [56] reported five charged isoforms for the MAb, HER2. Hunt et al. [56] demonstrated that charged variants could arise from cleavage of C-terminal lysine residues as well as deamidation of the molecule. Antibodies produced in mammalian cell culture tend to lack terminal sialic acid residues and, therefore, charge heterogeneity is not due to sialic acid. CZE was performed by Guzman et al. [68] to quantitate MAb anti-TAC using an uncoated capillary with 50 mM sodium borate at pH 8.3. Lee [69] published on the CE-SDS

analyses of animal immunoglobulins and murine monoclonal antibodies using the 'CE-SDS Protein Kit' from BioRad Laboratories, Hercules CA, USA. Under non-reduced conditions, murine monoclonal antibodies show a predominant peak with five to six apparent fragment peaks. The magnitude of the fragmentation is temperature, pH, and buffer dependent. Without heat treatment during the preparation of the SDS-antibody complexes, the fragmentation is almost non-existent. However, Lee observed several peaks near the expected migration time of an IgG with some murine monoclonal antibodies and suggested that this was due to their anomolous interaction with SDS. Therefore, for immunoglobulins, the preparation of samples for CE-SDS must be investigated before routine performance.

An early review by Kakehi and Honda [17] on the HPCE analysis of glycoproteins, glycopeptides, and glycoprotein-derived oligosaccharides presented the analysis of protein glycoforms and the mapping and confirmation of glycopeptides. This review also provided detailed information on the analysis of carbohydrate chains released from the peptide core, by direct detection and through derivatization. Subsequently, Pantazaki et al. [18] published a review on CE of recombinant glycoproteins. They described the main strategies that are available for the CE analysis of glycoproteins; CZE, cIEF, MEKC, and CE-SDS-NGS and described the practical problems that have been encountered in quality control. The strategy to overcome adsorption of proteins to the capillary wall and the lack of solubility of glycoproteins were discussed. Due to the advantages of simplicity, speed, and automation, CZE has been employed for the routine analysis of glycoform populations of intact recombinant glycoproteins and for monitoring their production. Pantazaki et al. [18] suggest that MEKC is an alternative method to CZE for purity control of recombinant glycoproteins. A recombinant basic glycoprotein, FG, was analyzed by CZE using a capillary with a dynamic coating of an amphipathic polymer [70]. This chimeric glycoprotein was resolved into two peaks; and the authors were not able to resolve this protein into additional glycoforms. CZE of native human IL-2 can resolve this glycoprotein into unsialylated, mono, and di-sialylated forms [71].

In the remainder of this review, more recent publications on the use of CE during product development of recombinant proteins will be discussed. Affinity CE will not be addressed although applications using this mode of CE are now beginning to be developed and used for the analysis of recombinant proteins.

1.5 Recent CE Literature

1.5.1 Protein Purification and Quantification

Jorgensen et al. [72] used the elution position in CZE of rhGH and its precursor biosynthetic human growth hormone (PrebhGH) to develop a rapid method for quantification of GH in E. coli cell pastes. These cell pastes can render conventional chromatographic columns unusable quite rapidly. Therefore, the use of CE is an attractive alternative. Jorgensen et al. [72] used hydrophobic C_{18} coated capillaries with biosynthetic hGH as an internal standard to quantify Pre-bhGH. The method was able to detect deamidated Pre-bhGH. By modifying the running buffer (150 mM Tricine, 7.5% methanol [v/v], pH 7.55) with zwitterionic surfactants and an organic modifier these authors detected another variant, trisulfide Pre-bhGH. His work was similar to that of McNerney et al. [57] except the fermentation media contained different variants of hGH. These differences could be attributed to the different E. coli hosts used for rhGH production. Arcelloni et al. [73] reported on a CZE method for the simultaneous and precise quantification of human insulin, proinsulin, and intermediate forms released in culture media by engineered cells. Culture supernatants were purified on Sep-Pak cartridges, recovered in an acetonitrile:trifluoroacetic acid mixture (60:40, v:v), concentrated, ultrafiltered, and analyzed by CZE. Protein recovery was $85 \pm 14\%$ and the method demonstrated a sensitivity of 0.5 nmol L^{-1} for quantifying insulin in the culture media. These authors were also able to compare the protein pattern released from engineered cells transfected with different human proinsulin constructs.

CZE has a tremendous advantage over traditional methods such as slab-gel IEF in analyzing basic proteins. Tran et al. [74] used reversed-charge CZE on amine-coated capillaries to monitor recombinant human platelet-derived growth factor (rPDGF) and showed that the method could be used to determine proteolytically

cleaved forms of rPDGF. rPDGF is secreted as a fully folded homodimeric protein consisting of two disulfide-linked chains. During fermentation, internal proteolysis yielded three forms; intact, singly, and doubly cleaved. This proteolysis also created new C-terminal sites and led to a very complex mixture of isoforms. RPHPLC methods did not resolve these isoforms. Separation of Enbrel® (rhuTNFR:Fc) isoforms by cIEF is presented in this monograph by Jochheim et al. [75]. Enbrel is a complex glycoprotein and details of optimization of the method and characterization of the peaks are provided. The method can resolve isoform species whose pIs differ by 0.05 pI units. This method was used to analyze glycoform patterns after process parameters were changed to determine consistency of the final product.

1.5.2 Protein Characterization

Vo et al. [76] developed a method for determining the gamma-carboxy glutamic acid (GLA) content of recombinant prothrombin since the GLA content was associated with the presence or absence of the calcium-dependent conformational changes required for prothrombinase function. They labeled a base hydrolyzate of the protein with FITC and separated the labeled amino acids using CZE with laser-induced fluorescence for detection. Both plasma and recombinant prothrombin contain 10 GLA residues per molecule of prothrombin.

HPCE was used to analyze glycosylated and non-glycosylated recombinant human granulocyte colony stimulating factor (G-CSF). Glycosylated G-CSF preparations contained HSA, added as a protein carrier. Glycosylated G-CSF could be separated into two distinct glycoform populations at pH 2.5 [77]. Differences in migration time and peak shape between the glycosylated and non-glycosylated variants could be demonstrated. The proteins were analyzed at two pH values. Whereas the non-glycosylated G-CSF migrated as a single sharp peak, the glycosylated material migrated as a doublet with the peaks being more defined at pH 2.5. CZE was able in all cases to differentiate the HSA peak from the active molecule [77]. Kinoshita et al. [78] describe the CZE analysis of sialoglycoproteins using DB-I capillaries. These authors analysed EPO, fetuin glycoforms, and α_1 acid-glycoprotein.

Recombinant human deoxyribonuclease (rhDNase) is an acidic, complex, glycoprotein consisting of 12–16 bands on slab-gel IEF, focusing between pI markers of 3.5 and 4.3. The charge heterogeneity is due to variable sialylation and phosphorylation of mannose structures. Calcium is required for the stability of rhDNase [79]. Felten et al. [80] had shown that the separation of rhDNase glycoforms was greatly improved by the addition of calcium ions to the CZE running buffer. The pH-dependent calcium binding effects on the electrophoretic separation were demonstrated at both acidic and basic pH. This resulted in a two-dimensional (pH 4.8 and 8.0) calcium aided analysis that achieved resolution of the charge microheterogeneity of rhDNase. In this monograph, Quan et al. [81] extend these studies by describing the effect of solution environments on the resolution of rhDNase variants by CZE. Hydrogen ion titration of the CZE buffer solution led to changes in electrophoretic mobility of rhDNase which was correlated to theoretical net protein charge. Divalent metal cations in the BGE were shown to associate with acidic rhDNase and modification of its electrophoretic mobility was related to the binding affinity of the metal ion interaction. Conditions that led to decreased electrophoretic mobility enhanced resolution of protein zones. These types of studies will continue to provide understanding of the CZE of complex protein molecules.

Characterization of recombinant human monoclonal antibodies have been described in considerable detail as there have been a number of antibodies approved for therapeutic use. There are several chapters in this monograph which describe the CE analyses of monoclonal antibodies [32, 34, 82]. Profiling of oligosaccharide-mediated microheterogeneity of a monoclonal antibody (MAb) was performed by Hoffstetter-Kuhn et al. [83]. These authors demonstrated that in CZE in phosphate buffer, only one peak is observed, and there was clear resolution from a glycine excipient. When the same MAb was analyzed using a borate-containing buffer, three peaks were detected which were attributed to the oligosaccharide microheterogeneity. Hoffstetter-Kuhn et al. [83] optimized the borate concentration to 150 mM and the best separation was obtained at pH 9.4. They obtained good precision for the peak areas (5.1–7.3% for the three peaks) and linearity was demonstrated up to a protein concentration of 0.1% (w/v) with a correlation coefficient > 0.999. This method was used for evaluating batch-to-batch consistency and for stability testing of the final formulated product. More recently, Santora et al. [84] have published the characterization of a monoclonal antibody to tumor necrosis factor-α. These authors used analytical techniques such as HPLC, CZE, gel methods, and mass spectrometry to perform their characterization. Compared to the gel-IEF pattern wherein three bands were observed, cIEF resolved four major peaks. Using carboxypeptidase B (CPB) digestion, the two earliest migrating peaks were attributed to the presence of lysine residues at the C-terminal of both and only one heavy chain, respectively, prior to the main peak (which contained no C-terminal lysine). Although the fourth peak was not detected on flat-bed IEF, four peaks were also observed on cation-exchange chromatography similar to the cIEF analysis. Using a combination of sialidase and CPB, these authors identified the contribution of sialic acid and C-terminal lysine residues to the charge heterogeneity observed. The acidic or fourth peak was due to sialic acid heterogeneity present on the Fc portion of the MAb molecule. Hagmann et al. [85] characterized a F(ab')$_2$ fragment obtained by pepsin cleavage of a murine MAb using ESI-MS, cIEF, high performance anion-exchange chromatography with pulsed amperometric detection, and LC-MS peptide mapping. cIEF of the F(ab')$_2$ showed five groups of peaks, the number of isoforms being reduced to three with sialidase treatment. cIEF under reducing conditions showed two peaks each of heavy chain (HC) and light chain origin. Sialidase treatment resulted in only one HC peak, and, therefore, the sialic acids were located only on the heavy chain of the F(ab')$_2$ fragment. The remaining charge heterogeneity was attributed to deamidation occurring on the light chain. This assumption was supported by ESI-MS of the reduced F(ab')$_2$ molecule. Tang et al. [86] proposed the use of cIEF for the routine analysis of recombinant immunoglobulins (rIgG). These investigators developed a cIEF method using a dimethyl siloxane-coated capillary (DB-1) and a separation matrix of 2% ampholytes in 0.4% methyl cellulose. Mobilization of focused bands was achieved by simultaneous application of low pressure and voltage. They showed good correlation to slab-gel IEF with comparable peak area

percent and pI determined for the individual components. For one MAb, an IEF method using slab-gel was not possible because it was a very basic rIgG (pI > 9.3). However, cIEF was able to resolve two charged isoforms for this basic rIgG. The DB-1 capillary was robust, no difference in the performance of the capillary was observed over 150 analyses. These authors demonstrated good inter-day precision, < 8% and < 3%, for rIgG peak areas and mobilization times, respectively.

1.5.3 Formulation Studies

McIntosh et al. [87] evaluated the use of CE for monitoring the effects of excipients on protein conformation using the model protein, ribonuclease (RNAse A) subjected to thermal denaturation. The effects of selected excipients on the thermal unfolding of RNAse A were then evaluated by adding sorbitol, sucrose, PEG 400, or 2-methyl-2,4-pentanediol to the electrophoretic run buffer at pH 2.3. Confirmatory experiments were performed using circular dichroism spectropolarimetry (CD). They determined that the observed changes in transition temperatures for RNAse A as a function of pH and selected excipients were similar by far UV, CD, and CE studies.

Sensitivity problems encountered in CZE separations can be reasonably overcome by coupling of CZE with on-line isotachophoretic (ITP) sample pre-concentration. Gysler et al. [88] investigated the utility of ITP-CZE and high performance size-exclusion chromatography (HPSEC) to analyse dimeric and monomeric rhIL-6. Dimer formation was evaluated over a wide range of pH and temperature. Incubation at pH 4.0 and 45 °C for 21 hours resulted in the reversible formation of a dimer as detected by HPSEC. The amount of dimer formation decreased as the pH was raised to pH 7.0. Nanomolar amounts of protein were required for these analyses. Therefore, alternative methods were sought to reduce sample consumption. ITP-CZE provided comparable results except that the dimer peak was resolved into two components. When the purified dimer peak separated by HPSEC was analyzed by ITP-CZE, the major peak was a doublet similar to that observed by ITP-CZE of the unfractionated rhIL-6. The peaks were analyzed with CZE-ESI-MS and all peaks had equivalent mass confirming that the

dimers were non-covalently linked and dimerized in different complexes. These complexes were only separable by the electrophoretic-based separation. These studies were extended to show that ITP-CZE-ESI-MS was able to identify fragments of IL-2 and could be performed on a small amount of starting material [89].

CZE with optimized temperature control for studying thermal denaturation of proteins at various pH values and conditions was reported by Rochu et al. [90, 91] using β-lactoglobulin as a model protein. These authors modified a commercial instrument to improve the control and measurement of temperature. They determined transition temperature, enthalpy change, and entropy change associated with thermal denaturation under various pH conditions. The thermodynamic parameters estimated compared well with parameters obtained by calorimetric measurements. Van't Hoff parameters were calculated through direct population determinations in a slow-time regime, and the stability curve (the temperature dependence of the free energy change of protein unfolding) was generated. They concluded from their studies that β-lactoglobulin exhibited complex pH and temperature association/dissociation behavior in its quaternary structure.

These authors stated that the technique is unique in its ability to estimate the heat capacity change and suggested that CZE is a powerful tool to study protein unfolding rapidly with minimum sample preparation. Stellwagen et al. [92] studied the effect of urea on the equilibrium conformation of cytochrome c in 20 mM Asp at 25 °C.

Cytochrome c was equilibrated with a series of solutions containing increasing amounts of urea. Electrophoresis was performed on aliquots injected into a capillary containing the same Asp/urea concentration and apparent mobility was determined. The measured apparent mobility values describe a transition between the folded and unfolded protein, with the unfolded protein being reduced in mobility. The authors suggest that this reduction reflects the larger frictional coefficient of the unfolded protein. Stellwagen et al. [92] proposed that this method could be used to examine the folding and stability of wild type, mutant, or post-translationally modified proteins.

Facchetti [93] performed stability studies on a recombinant protein using CE-SDS. He demonstrated that trace levels of aggregates of the recombinant product could be detected by this method and that it was more quantitative than slab gels. This author generated aggregates in his biopolymer solution by storage at −20 °C and at pH 7.0. From his data, Facchetti suggests that the CE-SDS method could be used as a limit test for aggregates. Salvi et al. [94] performed a careful optimization of the analytical parameters and injection procedure of CZE (pH, buffer concentration, capillary dimensions, temperature, field strength, and use of an internal standard) to develop a method which could monitor protein oxidation and screen for compounds which afforded antioxidant protection. Park et al. [95] described a CZE method for determining the purity and quantity of interferon alfacon-1 in formulations containing the interfering protein excipient, HSA. The CZE method was more precise than the ELISA method previously used to quantify interferon alfacon-1.

1.6 Quality Control and Validation

1.6.1 Purity

Dupin et al. [96] developed a CZE method to resolve and quantify variants of hGH such as a cleaved hGH, a histidine to glutamine 18 replacement, deamidated hGH, and succinylated hGH that were found in their recombinant preparations. This method used untreated fused-silica capillaries and a running buffer of 0.1 M diammonium hydrogen phosphate, pH 6.0. The method was remarkable in that hGH and these variants differed at most by 0.5 pI units and were baseline resolved, showing good specificity between the major variants. Linearity was demonstrated over the range of 0–1.17 mg mL^{-1} hGH. The precision determined with six replicate analyses performed over 4 days was < 8.2% RSD for the four variants ranging in percent of total hGH from 0.65–2.56%. The precision for the main peak at 0.58 mg mL^{-1} was 2.1%. A limit of detection of 0.03% was calculated for the variants which were considered to be impurities.

Silver-stained SDS-PAGE has been used extensively as a general purity test for lot release of protein pharmaceuticals. Hunt and Nashabeh [97] attained equivalent sensitivity using fluorophore-labeled proteins and enabled the replacement of

SDS-PAGE gels by CE-SDS non-gel sieving. Hunt and Nashabeh [97] used the amine-reactive fluorophore 5-TAMRA.SE to label a therapeutic MAb. They compared silver-stained SDS-PAGE to CE-NGS with LIF under non-reduced and reduced conditions showing comparability. Hunt and Nashabeh [97] mixed the MAb with 0.5, 1.0, and 5.0% (w/w) of a 22 kD recombinant protein and demonstrated that the method could detect the 22 kD protein as an impurity at the 0.5% (w/w) level.

1.6.2 Validation of CE Assays

Chen et al. [98] validated a CZE method for determining the purity of an anti-microbial peptide (protegrin IB-367). The method was validated with respect to accuracy, precision, linearity, range, limit of detection, specificity, and robustness. Schenerman and Bowen [34] describe the parameters of assay validation from the perspective of the regulatory requirements, and provide guidance on optimization strategies and protocol that are required in a quality control environment. These authors used a CE-SDS assay for a therapeutic monoclonal antibody to present examples of how the various parameters of validation would be approached. For example: specificity is demonstrated by showing the method can discriminate between the product and other proteins used in the process; linearity, which is required if the method will be used for purity assessment; range for quantitative applications; accuracy by demonstrating accurate molecular weights of the MAb fragments, compared to a reference method; precision; robustness; and sensitivity [34]. Hunt et al. [99] validated a cIEF method for determining the identity and charge distribution of a monoclonal antibody. The validation was performed according to ICH guidelines [100] so that the method could be used for product lot release. Hunt et al. [99] used ^{125}I-labeled antibody to demonstrate 99% recovery of the labeled material from the capillary. They determined that the repeatability and intermediate precision RSD for the migration time, peak area and peak area percent for the four major peaks ranged from 0.9–4.4%. Specificity was demonstrated by baseline resolution of the MAb main peak from product excipients, and from other recombinant produts. Linearity was demonstrated over

the concentration range of 2–356 µg mL^{-1}. Other parameters of validation that were performed were: focusing and mobilization voltage, capillary-to-capillary variability, alternate capillaries, and alternate instruments [99].

Hunt and Nashabeh [97] validated a CE-SDS method for determining purity of a recombinant monoclonal antibody. In their validation, they compared the CE-SDS to the current silver-stained SDS-PAGE method for reduced and non-reduced samples, to demonstrate accuracy of the observed size distribution profile. They optimized the dye-to-protein ratio evaluating the effects of labeling on rMAb aggregation. The sample preparation, which included heating of the protein in SDS, was also carefully evaluated as heating at 90 °C appeared to cause generation of fragments. These were key studies because this type of assay can be used to determine the level of the main peak (intact IgG) as a lot release requirement. The analysis of carbohydrate of a monoclonal antibody was described by Ma and Nashabeh [101]. This method was validated and used to demonstrate consistency of production with respect to carbohydrate. The accuracy of the assay was evaluated by recovery of known amounts of a carbohydrate standard, and quantitation was performed using a calibration curve.

1.7 Developments in CE-MS

CE-MS for on-line characterization is another area that if successful will further the acceptance and usage of CE in the biotechnology industry. Yeung et al. [102] used CE-MS to demonstrate the on-line identification of glycoform peaks separated by CE. Yeung et al. used a linear polyacrylamide coated capillary in conjunction with an acidic β-alanine buffer for the separation of proteins with N-linked high-mannose structures. For CE-MS studies, a VG Platform single quadrupole instrument with a VG ESI source was used with a coaxial CE-MS interface (Micromass, Manchester, UK). These authors used RNase B as a model protein for rhBMP-2 as these two proteins exhibit similar mass, isoelectric point and glycan structures. CE-UV and CE-MS profiles for RNase B were similar, and the MS results unequivocally identified each of the five glycoforms with high accuracy. However, the studies to identify the glycoforms of rhBMP-2 were more complex. Nine glycoforms of rhBMP-2, derived from three isoforms, can be resolved by CZE as described previously [46]. These isoforms can contain from 5 to 9 mannose residues. CE-MS experiments on the intact homodimer did not show good resolution in CE. However, a much improved separation was obtained on reduced and alkylated rhBMP-2 as detected by UV. Except for an aggregate peak, the masses detected corresponded to the extended, mature and pyroglutamate-containing form of the protein. Reconstructed ion chromatograms in ESI-MS were used to account for the microheterogeneity observed by CE-UV. Boss et al. [103] evaluated CZE-MS as a method for separation and characterization of complex peptides and glycopeptides of endoproteinase V8-digested rhEPO. The peptides were evaluated with and without mass detection to determine peptide purity. The peptide mass determined from the sequence was compared to the mass obtained from CZE-MS. Oligosaccharide analysis appeared to be straight forward, with glycosylation site occupancy determined qualitatively. Nonglycopeptides were resolved and analysed easily on-line and coverage of the protein sequence by CZE-MS was 98.2%. The capillary used was dynamically coated with polybrene in the presence of polyethylene glycol. It appears that the analysis of rhEPO peptides by CZE-MS was very successful. CE-MS for peptide mapping as a lot release assay may be the next achievement for the use of CE in the biotechnology industry. Liu et al. [104] described a system for rapid characterization of proteins with known sequences. The system consisted of CZE coupled to an electrospray ionization ion trap tandem mass spectrometer, via a sheath-flow interface. The procedure consists of CZE separation of peptide digests with MS-MS analysis of the peptides and a sequence database search to monitor for specific peptides. The proteins were identified at the low picomole level. This type of method can be useful for quality control purposes to monitor for specific peptides. Tong and Yates [105] review the current state of CE/MS/MS for protein identification in complex mixtures. CE/ESI-MS interfaces are described. Using a nanospray interface, Tong and Yates [105] demonstrate sub-femtomole detection limits of a protein digest. These studies indicate that CE-MS methods are close to being used on a routine basis.

1.8 References

[1] International Conference on Harmonization Topic Q6B: Guidance on Specifications: Test Procedures and Acceptance Critera for Biotechnological/Biological Products *Federal Register* **1999**, *64*, 44928.

[2] Harris, R.J.; Kabakoff, B.; Macchi, F.; Shen, F.J.; Kwong, M.; Andya, J.D.; Shire, S.J.; Bjork, B.; Totpal, K.; Chen, A.B. *Submitted for publication.*

[3] Boyd, P.N.; Lines, A.C.; Patel, A.K. *Molec. Immunol.* **1995**, *32*, 1311.

[4] Smedsrod, B.; Einarsson, M.; Pertoft, H. *Thromb. Haemostas.* **1988**, *59*, 480.

[5] Wittwer, A.J.; Howard, S.C.; Carr, L.S.; Harakas, N.K.; Feder, J.; Parekh, R.B.; Rudd, P.M.; Dwek, R.A.; Rademacher, T.W. *Biochemistry* **1989**, *28*, 7662.

[6] Dolnik, V. *Electrophoresis* **1997**, *18*, 2353.

[7] Dolnik, V. *Electrophoresis* **1999**, *20*, 3106.

[8] Righetti, P.J.; Bossi, A.; Olivieri, E.; Gelfi, C. *J. Biochem. Biophys. Methods* **1999**, *40*, 1.

[9] Wehr, T.; Zhu, M.; Rodriguez-Diaz, R. *Methods in Enzymol.* **1996**, *270*, 358.

[10] Rodriguez-Diaz, R.; Wehr, T.; Zhu, M.; *Electrophoresis* **1997** *18*: 2134.

[11] Guttman, A.; *Electrophoresis* **1996**, *17*, 1333.

[12] Strege, M.A.; Lagu, A.L. *J. Chromatogr. A* **1997**, *780*, 285.

[13] Denton, K.A.; Tate, S.A. *J. Chromatogr. A.* **1997**, *697*, 111.

[14] Strege, M.A.; Lagu, A.L. *Electrophoresis* **1997**, *18*, 2343.

[15] Lagu, A.L. *Electrophoresis* **1999**, *20*, 3145.

[16] Krull, I.S.; Liu, X.; Dai, J.; Gendreau, C.; Li, G. *J. Pharm. Biomed. Anal.* **1997** *16*, 377.

[17] Kakehi, A K.; Honda, S. *J. Chromatogr. A* **1996**, *720*, 377.

[18] Pantazaki, A.; Taverna, M.; Vidal-Madjar, C. *Anal. Chim. Acta* **1999**, *383*, 137.

[19] Wise, E.T.; Singh, N.; Hogan, B.L. *J. Chromatogr. A.* **1996**, *746*, 109.

[20] Bergmann, J.; Jaehde, U.; Schunack, W. *Electrophoresis* **1998**, *19*, 305.

[21] Wu, X.Z.; Hosaka, A.; Hobo, T. *Anal. Chem.* **1998**, *70*, 2081.

[22] Corradini, D. *J. Chromatogr. B* **1997**, *699*, 221.

[23] Allen, D.; Baffi, R.; Bausch, J.; Bongers, J.; Costello, M.; Dougherty, J.Jr.; Federici, M.; Garnick, R.; Peterson, S.; Riggins, R.; Sewerin, K.; Tuls, J. *Biologicals*, **1996**, *24*, 255.

[24] Janini, G.M.; Issaq, H.J. *This monograph.*

[25] Janini, G.M.; Metral, C.J.; Issaq, H.J.; Muschik, G.M. *J. Chromatogr. A* **1999**, *848*, 417.

[26] Offord, R.E. *Nature* **1966**, *211*, 591.

[27] Chen, J.; Fausnaugh-Pollitt, J. *This monograph.*

[28] Verzola, B.; Gelfi, C.; Righetti, P.G. *J. Chromatogr. A* **2000**, *868*, 85.

[29] Liu, X.; Sosic, Z.; Krull, I.S. *J. Chromatogr. A* **1996**, *735*, 165.

[30] Conti, M.; Galassi, M.; Bossi, A.; Righetti, P.G. *J. Chromatogr. A* **1997**, *757*, 237.

[31] Tran, N.T.; Taverna, M.; Chevalier, M.; Ferrier, D. *J. Chromatogr. A* **2000**, *866*, 121.

[32] Wehr, T.; Rodriguez-Diaz, R.; Zhu, M. *This monograph.*

[33] Ferguson, K.A. *Metab. Clin. Exp.* **1964**, *13*, 985.

[34] Schenerman, M.A.; Bowen, S.H. *This monograph.*

[35] Strege, M.A.; Lagu, A.L. *Anal. Biochem.* **1993**, *210*, 402.

[36] Eriksen, J.; Holm, K.A. *J. Cap Elec.* **1996**, *3*, 37.

[37] James, D.C.; Freedman, R.B.; Hoare, M.; Jenkins, J. *Anal. Biochem.* **1994**, *222*, 315.

[38] Kats, M.; Richberg, P.C.; Hughes, D.E. *Anal. Chem.* **1995**, *67*, 2943.

[39] Miksik, I.; Deyl, Z. *J. Chromatogr. A* **1999**, *852*, 325.

[40] Nashabeh, W.; Grieve, K.F.; Kirby, D.; Foret, F.; Karger, B.L. *Anal. Chem.* **1994**, *66*, 2148.

[41] Kalman, F.; Ma, S.; Fox, R.O.; Horvath, C. *J Chromatogr. A* **1995**, *705*, 135.

[42] Rudd, P.; Joao, H.C.; Coghill, E.; Fiten, P.; Saunders, M.R.; Opdenakker, G.; Dwek, R.A. *Biochemistry* **1994**, *33*, 17.

[43] Watson, E.; Yao, F. *Anal. Biochem.* **1993**, *210*, 389.

[44] Watson, E.; Yao, F. *J. Chromatogr. A* **1993**, *630*, 442.

[45] Klausen, N.K.; Kornfelt, T. *J. Chromatogr. A* **1995**, *718*, 195.

[46] Yim, K.; Abrams, J.; Hsu, A. *J Chromatogr. A* **1995**, *716*, 401.

[47] Nielsen, R.G.; Riggin, R.M.; Rickard, E.C. *J Chromatogr.* **1989**, *480*, 393.

[48] Cifuentes, A.; Poppe, H. *Electrophoresis* **1995**, *16*, 516.

[49] Chen, A.B.; Rickel, C.A.; Flanigan, A.; Hunt, G.; Moorhouse, K.G. *J Chromatogr. A* **1996** *744*, 279.

[50] Kubach, J.; Grimm, R. *J. Chromatogr. A* **1996**, *737*, 281.

[51] Thorne J.M.; Goetzinger, W.K.; Chen, A.B.; Moorhouse, K.G.; Karger, B.L. *J. Chromatogr. A* **1996**, *744*, 155.

[52] Frenz, J.F.; Wu, S.; Hancock, W.S. *J Chromatogr. A* **1989**, *480*, 379.

[53] Rush, R.S.; Derby, P.L.; Strickland, T.W.; Rohde, M.F. *Anal. Chem.* **1993**, *65*, 1834.

[54] Yowell, G.G.; Fazio, S.D.; Vivilecchia, R.V. *J. Chromatogr. A* **1993** *652*, 215.

[55] Moorhouse, K.G.; Rickel, C.A.; Chen, A.B. *Electrophoresis* **1996**, *17*, 423.

[56] Hunt, G.; Moorhouse, K.G.; Chen, A.B. *J. Chromatogr. A* **1996**, *744*, 295.

[57] McNerney, T.M.; Watson, S.K.; Sim, J.-H.; Bridenbaugh, R.L. *J. Chromatogr. A* **1996**, *744*, 223.

[58] Klyushnichenko, V.; Tishkov, V.; Kula, M.-R. *J. Biotechnology* **1997**, *58*, 187.

[59] Goldman, M.H.; James, D.C.; Rendall, M.; Ison, A.P.; Hoare, M. *AT Bull Biotech and Bioengineer.* **1998**, *60*, 596.

[60] Cifuentes, A.; Moreno-Arribas, M.V.; De Frutos, M.; Diez-Masa, J.C. *J. Chromatogr. A* **1999**, *830*, 453.

[61] Bietlot, H.P.; Girard, M. *J. Chromatogr. A* **1997**, *759*, 177.

[62] Girard, M.; Bietlot, H.P.; Cyr, T.D. *J. Chromatogr. A* **1997**, *772*, 235.

[63] Zhou, G.-H.; Luo, G.-A.; Zhou, Y.; Zhou, K.-Y.; Zhang, X.-D.; Huang, L.-Q. *Electrophoresis* **1998**, *19*, 2348.

[64] Farchaus, J.W.; Ribot, W.J.; Jendrek, S.; Little, S.F. *Appl. Environ. Micrbiol.* **1998**, *64*, 982.

[65] Roddy, T.R.; Molnar, T.E.; McKean, R.E.; Foley, J.P. *J. Chromatogr. B* **1997**, *695*, 49.

[66] Silverman, C.; Komar, M.; Shields, K; Diegnan, G.; Adamovics, J. *J. Liquid Chrom.* **1992,**15, 207.

[67] Pritchett, T.; Evangelista, R.A.; Chen, F.-T.A. *BiolTech* **1995,** 13, 1449.

[68] Guzman, N.A.; Moschera, J.; Iqbal, K.; Malick, A.W. *J. Chromatogr. A* **1992,** 608, 197.

[69] Lee, H.G. *J. Immunol Methods* **2000,** 234, 71

[70] Tsuji, K.; Little, R.J. *J. Chromatogr. A* **1992,** 594, 317.

[71] Knuver-Hopf, J.; Mohr, H. *J. Chromatogr. A* **1995,** 717, 71.

[72] Jorgensen, T.K.; Bagger, L.H.; Christiansen, J.; Johnsen, G.H.; Faarbaek, J.R.; Jorgensen, L.; Welinder, B.S. *J. Chromatogr. A* **1998,** 817, 205.

[73] Arcelloni, C.; Falqui, L.; Martinenghi, S.; Pontiroli, A.E.; Paroni, R. *Electrophoresis* **1998,**19, 1475.

[74] Tran, A.; Parker, H.; Levi, V.; Kunitani, M. *Anal. Chem.* **1998,** 70, 3809.

[75] Jochheim, C.; Novick, S.; Balland, A.; Mahan-Boyce, J.; Wang, W.-C.; Goetze, A.; Gombotz, W. *This Monograph.*

[76] Vo, H.C.; Britz-Mckibbin, P.; Chen, D.D.Y.; MacGillivray, R.T.A. *FEBS Lett.* **1999,** 445, 256.

[77] Somerville, L.E.; Douglas, A.J.; Irvine, A.E. *J. Chromatogr. B* **1999,** 732, 81.

[78] Kinoshita, M.; Murakami, E.; Oda, Y.; Funakubo, T.; Kawakami, D.; Kakehi, K.; Kawasaki, N.; Morimoto, K.; Hayakawa, T. *J. Chromatogr. A* **2000,** 866, 261.

[79] Oefner, C.; Suck, D. *J. Mol. Biol.* **1986,** 192, 605.

[80] Felten, C.; Quan, C.P.; Chen, A.B.; Canova-Davis, E.; Mc Nerney, T.; Goetzinger, W.K.; Karger, B.L. *J Chromatogr. A* **1999,** 853, 295.

[81] Quan, C.P.; Canova-Davis, E.; Chen, A.B. *This monograph.*

[82] Ma, S.; Nashabeh, W. *This monograph.*

[83] Hoffstetter-Kuhn, S.; Alt, G.; Kuhn, R. *Electrophoresis* **1996,** 17, 418.

[84] Santora, L.C.; Krull, I.S.; Grant, K. *Anal. Biochem.* **1999,** 275, 98.

[85] Hagmann, M.-L.; Kionka, C.; Schreiner, M.; Schwer, C. *J. Chromatogr. A* **1998,** 816, 49.

[86] Tang, S.; Nesta, D.P.; Maneri, L.R.; Anumula, K.R. *J. Pharm. and Biomed. Anal.* **1999,** 19, 569.

[87] McIntosh, K.A.; Charman, W.N.; Charman, S.A. *J. Pharm. and Biomed. Anal.* **1998,**16, 1097.

[88] Gysler, J.; Mazereeuw, M.; Helk, B.; Heitzmann, M.; Jaehde, U.; Schunack, W.; Tjaden, U.R.; van der Greef, J. *J. Chromatogr. A* **1999,** 841, 63.

[89] Gysler, J.; Helk, B.; Dambacher, S.; Tjaden, U.R.; van der Greef, J. *Pharm. Res.* **1999,**16, 695.

[90] Rochu, D.; Ducret, G.; Masson, P. *J. Chromatogr. A* **1999,** 838,157.

[91] Rochu, D.; Ducret, G.; Ribes, F.; Vanin, S.; Masson, P. *Electrophoresis* **1999,** 20, 1586.

[92] Stellwagen, E.; Gelfi, C.; Righetti, P.G. *J. Chromatogr. A* **1999,** 838, 131.

[93] Facchetti, I. *Biomedical Chromatogr.* **1998,**12, 138.

[94] Salvi, A.; Carrupt, P.-A.; Testa, B. *Helvetica Chimica Acta* **1999,** 82, 870.

[95] Park, S.S.; Sloey, C.J.; Chang, B.S *This monograph.*

[96] Dupin, P.; Galinou, F.; Bayol, A. *J Chromatogr. A* **1995,** 707, 396.

[97] Hunt, G.; Nashabeh, W. *Anal. Chem.* **1999,** 71, 2390.

[98] Chen, J.; Fausnaugh-Pollitt, J.; Gu, L. *J Chromatogr. A* **1999,** 853, 197

[99] Hunt, G.; Hotaling, T.; Chen, A.B. *J Chromatogr. A* **1998,** 800, 355.

[100] International Conference on Harmonization Topic Q2B: Guideline on the validation of analytical procedures: methodology *Federal Register* **1997,** 62, 27464.

[101] Ma, S.; Nashabeh, W. *Anal. Chem.* **1999,** 71, 5185.

[102] Yeung, B.; Porter, T.J.; Vath, J.E. *Anal Chem.* **1997,** 69, 2510.

[103] Boss, H.J.; Watson, D.B.; Rush, R.S. *Electrophoresis* **1998,**19, 2654.

[104] Liu, T.; Shao, X.-X.; Zeng, R.; Xia, Q.-C. 47 *J. Chromatogr. A* **1999,** 855, 695.

[105] Tong, W.; Yates, J.R. *This monograph.*

Selection of Buffers in Capillary Zone Electrophoresis: Application to Peptide and Protein Analysis

G. M. Janini / H. J. Issaq

SAIC Frederick, NCI-Frederick, P.O. Box B, Frederick, MD 21702, USA

2.1 Introduction

Separating proteins and peptides remains one of the most challenging tasks in analytical chemistry. In the past, conventional gel electrophoresis and high performance liquid chromatography were the most favored techniques for peptide and protein separation. Now capillary electrophoresis is being recognized as a powerful alternative or complimentart to these techniques because of the many advantages it offers, such as high speed, high resolution, automation, compatibility with physiological conditions, and real-time detection. CE is becoming increasingly prominent in bioanalytical research because it is ideally suited for handling small amounts of sample material. In preparing this chapter, our objective is to highlight the role of the buffer in CZE separation of peptides and proteins, to help the reader obtain an understanding of the underlying principles and practical consideration, and to provide useful groundwork for further study.

The buffer plays a central role in capillary zone electrophoresis. In analogy to chromatography, the buffer in CZE assumes the role of the mobile phase and the stationary phase. It is the semiconducting property of the buffer that allows the free-zone electrophoretic migration of analyte ions in an electric field. Electroendoosmosis, which is an integral part of CZE, is mainly driven by the residual charges on the inner wall of the capillary, and may be controlled by judicious selection of buffer composition. Addition of micelle-forming detergents, complexing agents, and chiral selectors to the running buffer imparts on the system a chromatographic separation mechanism (partitioning) in addition to separations based on charge-to-size ratio.

Traditional buffers used in conventional electrophoresis (slab gel or tube) may or may not be applicable to capillary electrophoresis. Factors, such as buffer ionic strength, running current, compatibility with other additives, and capillary-wall chemistry, and UV absorbance, that are not critical in conventional electrophoresis, have to be considered in choosing an appropriate buffer in CE. The selection of buffers for CGE, CIEF, CITF, and SDS_PAGE is beyond the scope of this chapter and will not be considered.

2.2 Buffer Parameters

2.2.1 Buffer Basics

A buffer, by definition, is a solution that maintains a constant pH by resisting changes in pH as a consequence of dilution or addition of small amounts of acids and bases. Buffer pH, type, concentration, and ionic strength influence many solution state properties, such as solubility, reaction rate, and mechanism. In CZE, solute migration velocity, separation, column efficiency, and peak shape are sensitive to changes in buffer characteristics. In particular, the pH is of crucial importance, creating the need for stringent buffer control. Buffer capacity, which is a quantitative measure of buffering ability, is another property that needs to be ad- dressed. The buffer capacity must be high enough such that the local pH and conductivity will not change as a result of sample introduction and migration across the capillary. Many buffer solutions are multicomponent systems, and some are mixtures of two or more buffering systems.

A simple buffering system is a solution that contains a weak acid and one of its salts (or a weak base and one of its salts). A rule of thumb is that the concentration of the two components should be within a factor of 10 of each other. To illustrate the elementary principles of buffer action, consider a solution containing a weak acid (HA) and its salt (MA). In what follows mathematical rigor is sacrificed for simplicity and clarity of presentation. The equilibria involved are as follows:

$$HA + H_2O \overset{K_a}{\rightleftharpoons} H_3O^+ + A^- \qquad (1a)$$

$$A^- + H_2O \overset{K_h}{\rightleftharpoons} OH^- + HA \qquad (1b)$$

where K_a is the ionization constant of HA and K_h is the hydrolysis constant of A^-. The concentration of the weak acid [HA] and its conjugate base [A^-] at equilibrium are calculated as follows:

$$[HA] = C_{HA} - [H_3O^+] + [OH^-] \qquad (2a)$$

$$[A^-] = C_{A^-} + [H_3O^+] - [OH^-] \qquad (2b)$$

where C_{HA} and C_{A^-} are the analytical concentrations of the acid and the salt respectively and [ ] denotes concentration.

In acidic solution [OH^-] is negligibly small and, except for very dilute solutions, [H_3O^+], to a very good degree of approxi-

0009-5893/00/02   18-09 $ 03.00/0

mation, is much smaller than both [HA] and [A$^-$]. Therefore, it could be assumed that

$$[HA] \cong C_{HA} \qquad (3a)$$

$$[A^-] \cong C_{A^-} \qquad (3b)$$

By substituting these values in the equilibrium expression presented by Eq. (1a) and rearranged, we obtain

$$[H_3O^+] = K_a \frac{C_{HA}}{C_{A^-}} \qquad (4)$$

By using the definitions, pH = $-\log$ [H$_3$O$^+$] and pK_a = $-\log K_a$, we get the expression that is well known as the Henderson-Hasselbach equation:

$$pH = pK_a + \log \frac{C_{A^-}}{C_{HA}} \qquad (5)$$

Equation (5) is used for calculating concentrations of HA and A$^-$ needed to prepare buffers of any desired pH value, provided that (a) the pK_a of the acid is accurately known, and (b) the assumptions Eqs. (3a) and (3b) are valid. Assumptions (3a) and (3b) fail if K_a is large ($K_a > 1 \times 10^{-3}$) or the concentration of either HA or A$^-$, or both, are small and comparable with [H$_3$O$^+$]. Almost all practical buffers (see Table I) have small K values and are prerpared at concentrations high enough to render assumptions (3a) and (3b) valid. Table I lists a compilation of weak acids, bases, and zwitterions that are commonly used as buffer material [1–5].

In contrast to unbuffered solutions, buffer solutions are resistant to changes in pH as a result of dilution and the addition of small amounts of acids and bases, as clearly demonstrated in Figures 1 and 2, respectively. Figure 1 shows the effect of dilution on the pH of buffered and unbuffered systems. Note that the pH of unbuffered systems (HCl and CH$_3$COOH) increases continuously with dilution, while the pH of the buffer (CH$_3$COOH + CH$_3$COONa) remains constant over a wide range of concentration. Figure 2 shows the effect of addition of small amounts of acids or bases to buffered and unbuffered solutions. Whereas the addition of small amounts of an acid or a base to an unbuffered solution drastically changes its pH, such additions result in negligible changes in the pH of buffered solutions. All calculations required to generate the data for the figures were according to Eqs. (1)–(5).

Buffer capacity, sometimes referred to as buffer value or buffer unit, is a quanti-

Table I. Selected CE Buffers

Compound	pK (25 °C)	Abs.[a] (mAU) (200/220 nm)
Oxalic acid	1.23 (pK_1) 4.19 (pK_2)	
Phosphoric acid	2.12 (pK_1) 7.21 (pK_2) 12.32 (pK_3)	100/50 sodium phosphate 1200/150 sodium phosphate
Iminodiacetic acid	2.33 (50 mM)[a] (pI)[b]	
Aspartic acid	2.77 (50 mM)[a] (pI)[b] 3.00 (50 mM)[a] (pI)[b]	
Malonic acid	2.9 (pK_1) 5.7 (pK_2)	
Citric acid	3.06 (pK_1) 4.74 (pK_2) 5.40 (pK_3)	> 3000/2600 sodium citrate
Formic acid	3.75	> 3000/2600 sodium formate
Succinic acid	4.19 (pK_1)	> 3000/> 3000 sodium
succinate	5.57 (pK_2)	
Acetic acid	4.74	2100/900 sodium acetate
MES {2-(N-morpholino)ethanesulfonic acid}	6.15	2100/190
ADA (N-2-acetamidoiminodiacetic acid)	6.60	
PIPES {piperzine-N,N'-bis (2-ethanesulfonic acid)}	6.80	
ACES (N-2-acetamido-2-aminoethanesulfonic acid)	6.90	
Imidazole	7.00	
MOPS {3-(N-morpholino) propanesulfonic acid}	7.20	
TES {2-{tris(hydroxylmethyl)methyl} amino}ethanesulfonic acid}	7.50	
HEPES (N-2-hydroxyethylpiperazine-N'-2'-ethanesulfonic acid)	7.55	
Hydrazine	7.99	
HEPPS (N-2-hydroxyethylpiperazine-N'-3-propanesulfonic acid)	8.00	
TRICINE {N-{tris-(hydroxymethyl)methyl} glycine}	8.15	
TRIS {tris-(hydroxymethyl)aminomethane}	8.30	1750/10
BICINE {N,N-bis(2-hydroxyethyl)glycine}	8.35	> 3000/> 3000
Glycylglycine	8.40	
Boric acid	9.24	300/50 sodium borate
Ammonia	9.26	
Ammonium carbonate	6.35 (pK_a, 20 °C) 10.33 (pK_b, 20 °C)	
CHES {2-(cyclohexylamino-ethanesulfonic acid)}	9.50	2600/700
Trimethylamine	9.87	2200/750 acetate
CAPS {3-(cyclohexylamino) propanesulfonic acid}	10.40	2700/740

a. Absorbance values at buffer concentration of 40 mM, 1 cm path length [3]. b. From [4].

tative measure of the buffering power of the buffer solution. It is defined as the number of equivalents of strong acid or base needed to cause a 1 unit change in the pH of 1 L of a buffer solution. Buffer capacity is directly proportional to the concentration of the species that constitute the buffer as well as their concentration ratio. The higher the concentrations and the closer to unity the concentration ratio the higher is the buffer capacity. Buffers are most effective within a 2-unit pH range bracketing the pK_a of the acid (buffer range pK_a – 1 to pK_a + 1; i.e., $10 \le C_{A^-}/C_{HA} \ge 0.1$). For example, an acetic acid-acetate buffer is most effective in stabilizing the pH of the solution in the pH range 3.74–5.74. At any given buffer concentra-

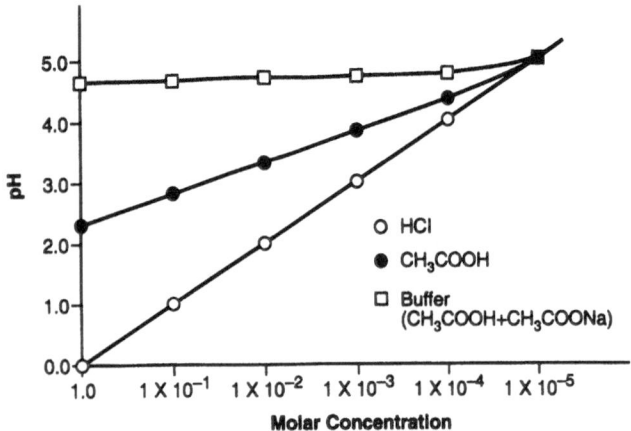

Figure 1. Effect of dilution on the pH of buffered and unbuffered acidic solutions. Initial concentrations: 1 M each; Temperature: 25 °C. K_a (CH$_3$COOH) = 1.85×10^{-5}. Reprinted with permission from [43].

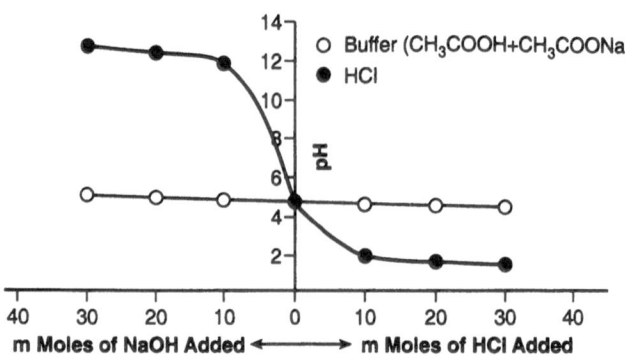

Figure 2. Effect of the addition of small amounts of acids or bases to buffered and unbuffered solutions. Initial concentrations: [CH$_3$COOH] = [CH$_3$COONa] = 0.5 M; pH = 4.74; [HCl] = 1.85×10^{-5}; pH = 4.74; K_a (CH$_3$COOH) = 1.85×10^{-5}. Reprinted with permission from [43].

Table II. Buffer Systems used in Protein Research

A. Commonly-used Ionic Salts Buffers	
Buffer	Buffer range
HCl/KCl[a]	1.0– 2.2
Glycine/HCl	2.2– 3.6
Na$_2$HPO$_4$/citric acid	2.6– 7.6
Citric acid/sodium citrate or NaOH	3.0– 6.2
Acetic acid/sodium acetate or NaOH	3.7– 5.6
NaH$_2$PO$_3$/Na$_2$HPO$_4$	5.8– 8.0
Bis. Tris/HCl	5.8– 7.2
Bis. Tris propane/HCl	6.3– 9.5
Triethanolamine hydrochloride/NaOH	6.8– 8.6
Tris/HCl	7.0– 9.0
Diethanolamine/HCl	8.0–10.0
Sodium borate/HCl	8.1– 9.0
Glycine/NaOH	8.6–10.6
Sodium carbonate/sodium bicarbonate	9.0–10.7
Sodium borate/NaOH	9.3–10.7
Na$_2$HPO$_4$/NaOH	11.0–11.9

B. Commonly-used Zwitterionic "Good" Buffers	
Buffer	Buffer range
MES/NaOH	5.5– 6.7
ADA/NaOH	6.0– 7.2
PIPES/NaOH	6.1– 7.5
ACES/NaOH	6.2– 7.5
MOPS/NaOH	6.5– 7.9
HEPES/NaOH	6.8– 8.2
TRICINE/HCl	7.4– 8.8
BICINE/HCl	7.6– 9.0
TAPS/NaOH	7.7– 9.1
CHES/NaOH	8.6–10.6
CAPS/NaOH	9.7–11.1

a) HCl/KCl has no buffering capacity. TAPS = N-Tris(hydroxymethyl)methyl-3-aminopropanesulphonic acid; others as defined in Table I.
Adapted from E.L.V. Harris and S. Angal, Protein Purification Methods, A Practical Approach, Oxford University Press, New York, USA, 1989, pp. 6–7, [11].

tion, its buffering capacity is maximum at pH = 4.74, and decreases symmetrically on both sides as pH is increased or decreased. The buffering power becomes extremely weak and ineffective when the pH is outside the range of $pK_a \pm 1$. It is, therefore, apparent that to prepare a buffer at a desired pH, the acid chosen should have a pK_a that is reasonably close to the desired pH. When selecting a buffer for a CZE application, buffer capacity cannot be overemphasized. At this stage of development, CZE still suffers from irreproducibility of migration times. One possible reason, among others, for this is the change of local pH of the running buffer inside the column as a result of sample introduction. This, however, can be avoided with careful selection of buffer systems with the right pK_a and concentration. One could point to numerous literature citations of CZE work conducted with buffers of minimal buffer capacity. Righetti et al. [4] highlighted this point with reference to an example from published literature.

The amounts of material needed to prepare a buffer of defined properties can be calculated. However, approximations in the equation used for buffer pH calculations and in the values of the weak acid or base dissociation constants give calculated pH values that are slightly different from measured values. Because of the laborious calculations involved, several authors have published tables and practical procedures for the preparation of buffers [1, 6–10]. Others have presented programs for calculating the amounts of material needed to prepare buffers of any specifications [11]. Table II gives a list of buffer preparations frequently used in protein electrophoresis [12].

2.2.2 Temperature and Joule's Heat Considerations

Temperature changes affect the equilibrium constants of buffer systems and, consequently, change the pH of the solutions. For acidic buffers, pH generally decreases with increasing temperature. Most buffers have $\Delta pK_a/\Delta T$ (change in pK_a per degree Kelvin) of about –0.01 [2]. A $\Delta pK_a/\Delta T$ value of –0.01 translates to a decrease of about 0.1 pH units for every 10 °C rise in temperature. Some buffers have higher $\Delta pKa/\Delta T$ values and, consequently, their pH can change significantly with temperature. For example, the pK_a

of Tris at 4 °C is 8.8, whereas at 20 °C it is 8.3. For this reason, buffers should be equilibrated at the temperature at which they will be used before making final pH adjustments [12]. This becomes critical if buffers are used at temperatures far removed from ambient [13].

More importantly is the control of the amount of heat generated inside the capillary and the facilitation of its dissipation such that it will not affect buffer pH, analyte thermal stability, and column efficiency. When a potential difference is applied across a buffer-filled capillary of finite resistance, energy is supplied to the system at a rate governed by the expression:

$$W = V \cdot I \quad (6)$$

where W is the rate of energy generation in Watt (Joule per second), and I is the current. Most of this energy is converted to heat which, if not efficiently dissipated, tends to (a) raise the buffer temperature, (b) cause convection currents, and (c) cause parabolic temperature variation across the column. The heat generated is dissipated into the column surroundings through the column wall. Minimizing heat generation and maximizing heat dissipation is critical in CZE instrument design because the presence of thermal gradients and other heat-related effects could adversely affect the very high efficiency of this technique. The amount of heat generated per unit time can be related to relevant CZE experimental parameters by substituting Ohm's law ($V = IR$) into Eq. (6):

$$W = V \cdot I = \frac{V^2}{R} \quad (7)$$

where R, the electrical resistance of the buffer is given by: $R = [L / (\Lambda CA)]$. In this relationship Λ is the molar conductivity of the buffer and A is the cross-sectional area of the capillary. A is equal to $\pi d^2/4$, where d is the capillary diameter and L is the capillary length.

Substituting for R in equation (7) gives:

$$W = \frac{\pi d^2 \Lambda C V^2}{4L} \quad (8)$$

Equation (8) reveals direct proportionality between the amount of heat generated inside a capillary and the molar conductivity (i. e., buffer type), and buffer concentration.

For a typical CZE experiment with $V = 25$ kV, $C = 10$ mM, $\Lambda = 126.45$ S cm² mol⁻¹ (NaCl, from Table III), $L = 50$ cm,

Table III. Ionic Mobilities and Limiting Ionic Conductivities of Selected Ions in Water at 25 °C

Ion	Ionic mobility (10^4 cm² s⁻¹ V⁻¹)	Limiting ionic conductivity (S cm² mol⁻¹)
H^+	36.23	349.6
Li^+	4.01	38.7
Na^+	5.19	50.10
K^+	7.62	73.50
Rb^+	7.92	77.8
Cs^+		77.2
Zn^{2+}	5.47	105.6
OH^-	20.64	199.1
Cl^-	7.91	76.35
Br^-	8.09	78.1
I^-	7.96	76.8
CH_3COO^-	4.24	40.9
$(COO)_2^{2-}$		148.2

Source: [14], Tables 27.1 and 27.2.

and $d = 75$ µm, the quantity of heat generated per second is estimated to be 0.7 W (i. e., 1.4 W/m). If this quantity of heat is absorbed by the small amount of buffer inside the column (2.2 µL), it will result in a considerable rise in temperature. Clearly, efficient heat dissipation is critical for the success of CZE experiments.

It is estimated that the temperature coefficients of mobility are nearly the same for all ions in a given solvent and are of the same order of magnitude as the temperature coefficient of viscosity [14]. Therefore, it is expected that the mobility increases by about 2% per kelvin rise in temperature in the neighborhood of 25 °C. Thermal gradients across the column cause the ions in the center of the column to move faster than the ions closer to the wall, resulting in zone-spreading and distortion. Wieme [15] gave the following expression for the temperature profile inside the column:

$$t_c - t_r = \left(\frac{0.239 W}{4 \kappa}\right) r^2 \quad (9)$$

where t_c = temperature in the center of the column; r = distance from center; κ = thermal conductivity (1.45×10^{-3} cal S⁻¹ K⁻¹ for water); and w = power per cubic centimeter. From this equation, a value of 0.18 °C is calculated for ($t_c - t_r$) for the CZE experimental setup described earlier in this section. This translates to a 0.4% difference in mobility between ions in the center and ions close to the column wall, a value that does not significantly affect column efficiency.

The importance of minimizing heat generation by careful selection of the buffer type and concentration and other CZE experimental parameters, and the importance of efficient heat dissipation by careful instrumental design, cannot be over-emphasized.

2.2.3 Buffer pH

Although the pH does not directly enter into any of the fundamental equations of CZE, it is, perhaps, one of the most important parameters that can be manipulated to optimize CZE separations. The pH of the buffer influences the ionic state of the capillary wall, thereby altering both the electroosmotic mobility and solute-capillary wall interactions. The charge of the capillary wall surface and the zeta potential are influenced by pH [16–18]. This is particularly true with fused-silica capillaries, owing to the ionization of acidic silanol groups. The ionization constant of the Si-OH groups on fused-silica surfaces is not precisely determined, but it is estimated to be about 1×10^{-3} [18]. As the negative charge is built up at the surface with increasing pH, electroosmotic mobility increases [16–19]. The magnitude of change in electroosmotic mobility depends on the surface chemistry of the capillary, and the presence of buffer additives (such as inorganic salts, organic solvents, surfactants, and polymers) that dynamically adsorb to the wall surface [19–23].

More importantly, the ionic equilibrium state of the analytes is influenced by pH. Thus, changes in pH result in changes in analyte mobility, and small differences in analyte pK values can be the basis for the separation of peptides and proteins of closely related pI values. Although intermediate pH values (close to the pI) have been used, the majority of literature citations report separations of peptides and proteins at the two extreme ends, acidic buffer (pH ≪ pI) and basic buffers (pH ≫ pI). At high pH (pH approximately 10), both the peptide or protein and the capillary wall are negatively charged resulting in minimum wall adsorption due to a charge repulsion effect, and at low pH (pH approximately 2), the peptide or protein is positively charged while the capillary wall is neutral minimizing electrostatic interactions. Each of these pH extremes has its own particular advantages and disadvantages which will be covered later in this review.

2.2.4 Buffer Type and Concentration

Buffer Type

In a series of articles, Issaq and co-workers [24–27] used probe solutes and electroosmotic markers to highlight the effect of the proper selection of buffer type on electroosmotic flow, electrophoretic mobility, interactions of analyte with capillary wall and current generation and Joule's heating. Their results show that electroosmotic flow and analyte electrophoretic mobility decrease as the cation size of alkali salt buffers increase from Lithium to Cesium ($Li^+ > Na^+ > K^+ > Rb^+ > Cs^+$). The current generated using 0.1 M alkali acetate buffers at the same voltage under the same experimental conditions increased in the order $Li^+ > Na^+ > K^+ > Rb^+ > Cs^+$ [26]. In another study the same group found that buffer anion type affects not only electroosmotic flow and analyte electrophoretic mobility but also selectivity and resolution [27]. The anions compared in the study were acetate, phosphate, borate, citrate and carbonate. It was also found that when tested at the same applied voltage, the different anions generated widely different currents. Carbonate produced the lowest current while citrate produced the highest. Tran et al. [28] conducted similar experiments and obtained corroborative results. They reported that a phosphate buffer gave better separation of a human erythropoietin than did sulfate or acetate buffers, and they attributed this to the strong interaction of the phosphate ions with the capillary surface. In other related studies Green and Jorgenson [29] found that elevated concentrations of different alkali salts (LiCl, NaCl, KCl, and CsCl) reduced the adsorption of the protein, lysozyme, to the capillary wall, in the order $Cs^+ > K^+ > Na^+ > Li^+$, cesium being the most effective and Li the least effective. Thus, more attention should be paid to the proper selection of buffer cations, anions and their counter ions.

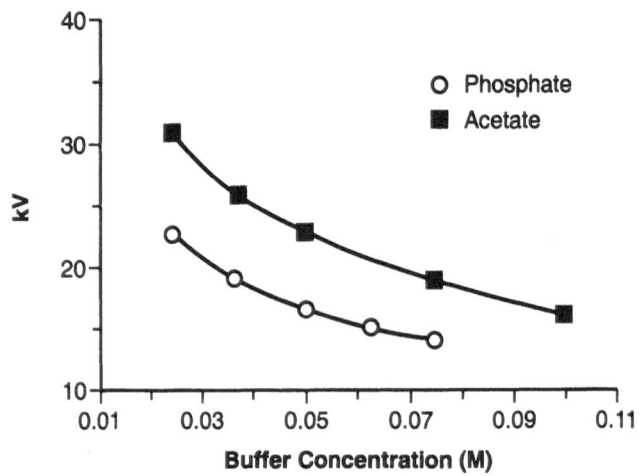

Figure 3. Maximum voltage versus buffer concentration. Column: 50 cm × 75 μm bare fused-silica; Instrument: Beckman Model P/ACE System 2000; Reprinted with permission from [24].

Buffer Concentration and Ionic Strength

Several researchers [30–33] have reported data on the effect of buffer concentration on mobility. Altria and Simpson [30, 31] reported that the mobility is inversely proportional to concentration and that a plot of the logarithm of concentration versus mobility is linear. Bruin et al. [33] studied overall mobility versus buffer concentration and concluded that the relation is linear. Nashabeh and El Rassi [32] also presented data on the effect of buffer concentration on mobility. A theoretical treatment of this subject was conducted by Issaq et al. [24] who derived an expression for the dependence of electrophoretic mobility (μ_{ef}) on buffer concentration:

$$\mu_{ef} \cong \frac{e}{3 \times 10^7 \mid Z \mid \eta \sqrt{C}} \qquad (10)$$

where e, Z, and C are the total excess charge in solution per unit area, the number of valence electrons, and the concentration of the buffer, respectively. The dependence of μ_{ef} on the reciprocal of the square root of concentration is in qualitative agreement with literature [15]. In their study, Issaq et al. [24] provided experimental verification of the dependence of μ_{ef} on buffer concentration as given in Eq. (10). They measured μ_{ef} for dansyl alanine and μ_{eo}, with mesityl oxide as a marker, at different concentrations using an acetate and a phosphate buffer system. All experiments were conducted at an electric field strength of 90 V/cm to minimize Joule-heating complications. The data was plotted as mobility versus the reciprocal of the square root of buffer concentration. All plots were linear, with correlation

coefficients in excess of 0.995. As the buffer concentration was increased, the electrophoretic and electroosmotic mobilities were both reduced. Equation (10) suggests a direct relation between solute migration time and the square root of concentration such that the migration time will double for every fourfold increase in buffer concentration. The effect of buffer concentration on column efficiency and resolution is not as dramatic as that of applied voltage; however, the consensus is that resolution increases with increasing buffer concentration [16, 24, 30–36].

A practical procedure for estimating the maximum concentration a buffer can have for a given column at a given voltage was suggested [24]. Figure 3 provides a plot of maximum voltage versus buffer concentration for sodium phosphate and sodium acetate buffers for the system defined in the figure caption [24]. The data was collected as follows. For each buffer concentration, an Ohms law plot (current vs. voltage) was generated. From each plot the maximum voltage at which the plot starts to deviate from linearity was recorded as a point in Figure 3. Deviation from linearity indicates a change in the electrical resistance of the buffer as a result of increase in temperature inside the capillary. The temperature inside the capillary starts to rise when the amount of Joule's heat generated as a result of high voltage and/or concentration overwhelms the instrument's mechanism of heat dissipation. For example, Figure 3 shows that when a 0.05 M sodium phosphate buffer is used, the onset of excessive Joule's heating is around 18 kV. In contrast, 25 kV may be applied to a 0.05 M sodium acetate buffer [24]. The lower the molar conductivity of the buffer is in CZE, the higher the voltage that may be applied [24, 26, 27].

In an interesting series of papers, Cross and co-workers [37, 38] reported CZE applications with buffers at concentrations higher than the upper limit described above. They reported that between the onset of Joule heating and the catastrophic effect of excessive heat production, there is a range of high buffer concentration in which analyte resolution is enhanced [37]. They developed theoretical treatments and presented experimental results in support of their argument that CZE at high buffer concentrations holds considerable promise for difficult separations, especially for ionizable solutes where the variation in the degree of ionization from analyte to analyte provides an additional source of selectivity [37, 38].

2.2.5 Isoelectric Buffers

An interesting class of buffers, namely, acidic isoelectric (amphoteric) buffers, has been promoted by Righetti and co-workers [4, 5] for the separation of peptides, proteins and oligonucleotides. The main advantage of these buffers is their much lower conductivity compared to ionic salt buffers. This allows the application of high buffer concentrations, high operating voltages and wide bore capillaries, resulting in shorter analysis time and, consequently, high resolution due to minimal diffusion-driven peak spreading. Isoelectric buffers have been used for various peptide and protein applications including peptide mapping, separation of different classes of proteins such as wheat proteins and denatured globins, and study of protein folding. On the negative side, these buffers exhibit higher UV absorptivity at 200–220 nm, resulting in reduced detection sensitivity compared to most conventional ionic salt buffers. Also, not all amphoteres have enough buffering capacity at their isoelectric point to be useful as a running buffer in CZE. An excellent theoretical treatment of the subject and practical guidelines for the selection of functional isoelectric buffers is provided by Righetti et al. [4]. Readers interested in the use of isoelectric buffers for the separation of peptides and proteins may consult recent reviews of the subject [4, 5].

2.2.6 Practical Considerations

Buffers Preparation

Most literature citations report buffer concentration and pH (e. g., 50 mM phosphate, pH 7.2) without giving full details of how buffers are made. As was discussed earlier, the counter ion and buffer ionic strength do affect electrophoretic migration and electroosmotic flow. To illustrate, a 50 mM acetate buffer may, for example, be made by mixing 50 mM lithium, sodium or potassium acetate, and 50 mM acetic acid in appropriate proportions to obtain the desired pH. Alternatively, ammonium, sodium, or potassium hydroxide can be added to 50 mM acetic acid to obtain the desired pH, which would result in buffers with different counter ions and ionic strength. Admittedly, pH of the buffer is more critical than the ionic strength or the nature of the counter ion; however, complete buffer characterization is necessary for reproducible results. The same buffer, prepared in different ways, can have dramatic effect on CZE properties [25–27, 39].

Buffers and Capillary Wall Chemistry

Capillary surface chemistry is a factor that has to be considered in buffer selection. Capillaries used to conduct CZE separations are made of either bare fused-silica with surface silanol groups exposed to the buffer media or coated fused-silica where the silanol groups are chemically modified with neutral, positively or negatively charged functionalities. Some buffers and buffer additives are selected for a certain CZE application because of their beneficial interaction with the capillary wall, and others are avoided because of their undesirable interaction with the analytes or capillary wall. McCormick [17] reported that phosphate ions interact with surface silanol groups at low pH, and he used this to an advantage for the high resolution of peptides. Primary amine-containing buffers can have a strong interaction with the wall, as was demonstrated by Chiesa and Horvath [40] who used a triethylammonium phosphate buffer at low pH to control and reverse electroosmotic flow. The same buffer additive affects the outcome of a CZE separation differently depending on the type of column used in the CZE experiment. For example, it is known that CTAB adsorbs to bare fused-silica surfaces and reverses EOF [41]. CTAB similarly adsorbs to C_8-coated

columns [19], but it does not adsorb to the surface of 10% linear polyacrylamide-coated columns, and the flow is not reversed resulting in a completely different profile for the same sample [42].

Buffers and Analytes

Possible interactions of buffer ions with analytes have to be considered. To illustrate, Tran et al. [28] have shown that the migration behavior of human erythropoetin is a function of buffer type, and Moring [3] reported that the use of sodium citrate in place of sodium acetate buffer resulted in significant difference in the migration pattern of basic peptides as a result of analyte-buffer interactions.

Miscellaneous

- CHAPS and other detergent-like additives form micelles in buffer solutions imparting chromatographic mechanisms to the separation of analytes in addition to differential electrophoretic migration.

- Citrate buffer, a strong chelator, forms strong ion pairs with some analytes such as basic peptides [3].

- The highest detection sensitivity for peptides and proteins is achieved in the wavelength range 195–210 nm, due to the absorbance of the peptide bond, which necessitates the selection of buffers with the least absorbance in this range.

- Other practical factors that influence the selection of buffers were discussed by Gueffroy [2]. These include stability to oxidation and light, complexation, and metal ion-binding properties.

2.3 Buffers Used for the Separation of Peptides and Proteins

2.3.1 Buffer Systems

The majority of books published so far on CZE have chapters that deal with peptide and protein separations. A survey of the literature reveals that a wide selection of buffers at different pH values, with and without additives, in bare- and coated fused-silica capillaries were used in the separation of peptides and proteins. Various aspects of this subject have been addressed in two reviews devoted to the selection of buffers in CZE [3, 43], and in many other general reviews on the CZE of peptides [45–47, and references therein] and proteins [48, 49, and references therein].

In this section we will highlight the various strategies used for the selection of buffer systems for peptide and protein separations with examples from recent literature. As mentioned in an earlier section, the choice of capillary material and capillary wall chemistry has a direct bearing on the proper selection of buffer constituents and pH. More often than not, buffer systems, pH and specific additives are selected in order to minimize the interaction of peptide and protein analytes with the capillary wall. The overwhelming majority of CZE experiments are conducted in fused-silica capillaries. At pH ≥ 2, the weakly acidic silanol groups on the surface of fused-silica start to ionize and form negatively charged groups. As pH is increased, the negative charge density on the capillary wall increases to a maximum at around pH = 10, where the silanol groups are completely ionized. The solutions proposed and used to minimize protein-wall adsorption tendencies include the selection of buffers at extreme pH (pH < 3 or pH > 8), use of buffer additives, and the use of coated capillaries. At pH < 3, the silanols on the capillary wall are mostly protonated and the proteins bear a net positive charge. As such, proteins are less likely to adsorb on a neutral surface compared to a negatively-charged surface. Undesirable adsorption artifacts are even further minimized if the column is coated, either covalently or dynamically with neutral hydrophillic polymers. The closest to a universal buffer for peptide and protein separation is a 50–100 mM phosphoric acid adjusted to pH in the range 2.2–3.2 with triethylamine [42, 45–47]. At this pH all peptides and proteins are positively charged and migrate in the same directions towards the detector, located at the cathodic end of the capillary. Compared to other buffering material, such as carboxylic acids and isoelectric amphoteres, phosphoric acid is much more light transparent at 200 nm, thus permitting higher detector sensitivity. Triethylamine, which is positively charged at this pH, helps cover any residual adsorption sites on the capillary wall, allowing the analytes to migrate as a discrete sharp zone. Bare silica capillaries can be used with this buffer if a viscous neutral polymer is also added to the buffer to further reduce electroosmotic flow. However, from our experience with peptide

mapping [42] and protein analysis [50], we favor the use of polyacrylamide-coated columns. Our laboratory routinely analyzes for proteins and peptides used in cancer and AIDS research. Our in-house polyacrylamide-coated columns that are dedicated to this application offer reproducible migration times to within ± 1% over months of continuous operation.

Peptides and proteins are also successfully analyzed in bare silica columns with extremely basic buffers, where the capillary wall is negatively charged. Peptides and proteins also carry a net negative charge at pH values greater than their pI's. Most proteins have pI < 8 and are, therefore, analyzable in bare silica capillaries with buffers in the pH range of 8–9.5 [45, 48, 49]. Under these conditions, the peptides and proteins are prevented from adsorption to the wall by electrostatic repulsion of like charges.

In rare occasions, buffer pH as high as 11 [51, 52] or as low as 1.8 [53] are employed for the separation of peptides and proteins. However, buffers with pH ≥ 11 or < 2 are not recommended because the large abundance of the highly mobile OH$^-$ ion at pH ≥ 11 and H$^+$ ion at pH < 2 generate excessive current and Joule heating. Also, the buffering power of electrolyte is poor or non-existent at these very extreme pH values. Cysteic acid, an isoelectric buffer, appears to be an exception. At a concentration of 200 mM, it registers a pI of 1.8 and exhibits a strong buffering power, allowing its use for peptide analysis at pH < 2. However, hexafluoro-2-propanol has to be added (up to 30% v/v) in order to bring the operating current and Joule heating down to managible levels [53].

The use of highly acidic or highly basic buffers has been utilized with a great deal of success for most peptide analysis applications, but the same thing could not be said about proteins. When confronted with the analysis of unknown proteins, a researcher may initially try extreme acidic or basic buffer conditions, but success cannot be guaranteed. Conditions that are ideal for the analysis of a particular protein may fail when applied to other proteins. There is always the risk of protein denaturation at pH extremes, and closely-related proteins may not be adequately separated at pH extremes. Also, many applications require buffers with pH where proteins are present in their native states. In an interesting review article, Corradini [54] provided a list of about 150 buffer systems that were used for protein

separations. We examined the list and sorted it with respect to pH. There were 63 applications at pH (> 8), 53 at pH (2–5.9) and only 33 applications at close to neutral pH (6–8).

2.3.2 Buffer Additives

Miscellaneous

Selectivity and resolution in CZE of peptides and proteins can be enhanced by incorporating appropriate additives into the buffer system. In general, buffer additives are added to modulate the electrophoretic mobility of the peptide or protein analytes, control electroosmotic flow and reduce analyte adsorption to the capillary wall. In 1997, Corradini published a comprehensive review of buffer additives for protein separations [54]. The review gives a list (Appendix A, [54]) of over 50 additives that have been used for protein analysis. The most frequently used additives are mono- and di-amines. The list also includes polyamines, neutral polymers, ionic salts, zwitterionics, ionic, non-ionic and zwitterionic surfactants, ion-pairing agents and urea. Readers are referred to [54] for comprehensive review of this subject. Other additives that were reported for protein separation, but were not covered in Corradini's review, include SDS, cyclodextrins and organic solvents. Organic solvents, such as methanol, ethanol, trifluoroethanol, isopropanol, and acetonitrile, have been used as buffer additive in the CZE separation of peptides and proteins [55, and references therein]. More recently, aqueous-organic buffers were used for the optimization of the separation of nine native forms of gonadotropen-releasing hormone decapeptides [56]. The separation was optimized by varying the percentage of acetonitrile, methanol, and isopropanol in the buffer [56]. The separation of alcohol-soluble proteins was facilitated by the addition of 20% acetonitrile to the running buffer [57], and 2,2,2-trifluoroethanol-aqueous buffers were shown to have advantages over unmodified aqueous buffers for the separation of peptides [58]. Cyclodextrins were also used as buffer additives for the separation of peptides and proteins [59–62]. Neutral, anionic, and cationic derivatives of beta-cyclodextrin were added to the buffer in an attempt to increase the resolution of peptides and proteins [59], and modified beta-cyclodextrins were used, to varying degrees of success, for the enhancement of the resolution of acidic as well as basic proteins [60–62]. A larger volume of literature has been devoted to the use of SDS as a buffer additive for the separation of peptides and proteins and, therefore, it will be discussed in a separate section below.

SDS

The use of SDS as well as other surfactants in non-sieving buffers is referred to as micellar electrokinetic chromatography. This technique has been routinely employed for the separation of peptides. Proteins, unlike peptides, are too large to effectively partition into micelles. Nevertheless, MEKC has been reportedly used for the separation of different classes of proteins.

Most peptides exhibit differences in charge-to-size ratio and are separated by CZE. For those that are not, MEKC offers a recourse. MEKC has been exploited for the separation of closely-related peptides that differ in the hydrophobicity of the amino acid residues. A few examples that demonstrate the power of MEKC are given below. MEKC was used for the separation of a synthetic set of heptadecapeptides that differ in a single amino acid using either the nonionic Tween 20 or the cationic CTAB [63]. A set of decapeptide angiotensin I variants were also separated using MEKC with the nonionic surfactant, Triton X-100 [64]. Another set of octapeptide angiotensin II analogs, differing in a single amino acid, have been separated using the nonionic surfactants Tween 20 [65, 66] and sucrose monododecanoate [67] at low pH, and were also separated using cationic surfactants at a neutral pH and anionic surfactants at high pH [68]. The separation of eight neurohypophyseal nonapeptide analogs was attempted using the cationic surfactant CHAPS, the anionic surfactant SDS, and the neutral surfactant Triton X-100 [69]. The zwitterionic surfactant PAPS has been reported to be effective in the separation of a few polymyxin decapeptide analogs [70]. Three closely related variants of 13-mer neurotensin analogs were separated using sucrose monododecanoate [67]. Longer peptides of 16- to 18-mers of synthetic basic neuropeptide Y analogs were separated using the cationic surfactant CTAC [69]. MEKC was also utilized for the separation of two 22-mer motilin variants using neutral surfactant Tween

20 or CTAB with 5% acetonitrile [71]. Finally, four variants of insulin from different species were separated using CTAB and 5% acetonitrile, and two 70-mer insulin-like growth factor I variants differing in a single amino acid were separated using the zwitterionic surfactant DAPS with 15% acetonitrile [72].

On the other hand, high molecular weight proteins are too large to fit and partition into a micelle, but there are several examples of improved protein separation using MEKC. Enhanced separation of proteins is most likely effected by interaction (hydrophobic, polar, electrostatic) of the protein with the outside surface of the micelle and/or with individual surfactant molecules. Model proteins were separated using SDS and CTAC [73]. MEKC with SDS was also used for the separation of recombinant proteins [73, 74], plasma apolipoprotein [75], antibodies [76, 77], cheese proteins [78], and glycoproteins [79]. The addition of organic modifiers to MEKC buffers is reported to improve the separation of large peptides [71] and proteins [80].

2.4 Acknowledgements

This project has been funded in whole or in part with Federal funds from the National Cancer Institute, National Institutes of Health, under Contract No. NO1-CO-56000.

By acceptance of this article, the publisher or recipient acknowledges the right of the U.S. Government to retain a nonexclusive, royalty-free license and to any copyright covering the article. The content of this publication does not necessarily reflect the views or policies of the Department of Health and Human Services, nor does mention of trade names, commercial products, or organizations imply endorsement by the U.S. Government.

2.5 References

[1] Long, C. Ed., *Biochemist's Handbook*, Van Nostrand, Princeton, New Jersey **1961**.

[2] Gueffroy, D.E. *"Buffers, A Guide for the Preparation and Use of Buffers in Biological Systems,"* Monograph, Behring Diagnostics, Doc. No. 4140–285, La Jolla, California **1985**.

[3] Moring, S.E. in *Capillary Electrophoresis in Analytical Biotechnology*, P. G. Righetti, Ed., CRC Press, Boca Raton, **1996** Chap. 2.

[4] Righetti, P.G.; Gelfi, C.; Perego, M.; Stoyanov, A.V.; Bossi, A. *Electrophoresis*, **1997**, *18*, 2145.

[5] Righetti, P.G.; Bossi, A.; Olivieri, E.; Gelfi, C. *J. Biochem. Biophys. Methods*, **1999**, *40*, 1, and references therein.

[6] Colowick, S.P.; Kaplan, N.O. Eds., *Methods in Enzymology*, Vol. 1, Academic Press, New York, **1955**.

[7] Bates, R.G. *Ann. N.Y. Acad. Sci.*, **1961**, *92*, 341.

[8] Elving, P.J.; Markowitz, J.M.; Rosenthanl, I. *Anal. Chem.*, **1956**, *28*, 1179.

[9] Good, N.E.; Winget, G.D.; Winter, W.; Connolly, T.N.; Izawa, S.; Singh, R.M.M. *Biochemistry*, **1966**, *5*, 467.

[10] Perrin, D.D. *Aust. J. Chem.* **1963**, *16*, 572.

[11] Lambert, W.J. *J. Chem. Educ.*, **1990**, *67*, 105.

[12] Harris, E.L.V.; Angal, S. *Protein Purification Methods, A Practical Approach*, Oxford University Press, New York, USA, **1989** pp. 6–7.

[13] Ma, S.; Horvath, C. *J. Chromatogr. A* **1998**, *825*, 55.

[14] Atkins, P.W. *Physical Chemistry*, 3rd Ed., Oxford Unviersity Press, Oxford, **1986**, Chap. 17.

[15] Wieme, R.J. in *Chromatography, A Laboratory Handbook of Chromatographic and Electrophoretic Methods*, 3rd Ed., E. Heftman, Ed., Van Nostran Reinhold, New York, **1975**, Chap. 10.

[16] Lukacs, K.D.; Jorgenson, J.W. *J. High Resolut. Chromatogr.* **1985**, *8*, 407.

[17] McCormick, R.M. *Anal. Chem.* **1988**, *60*, 2322.

[18] Hayes, M.A.; Ewing, A.G. *Anal. Chem.* **1992**, *64*, 512.

[19] Janini, G.M.; Chan, K.C.; Barnes, J.A.; Muschik, G.M.; Issaq, H.J. *J. Chromatogr.* **1993**, *653*, 321.

[20] Terabe, S.; Ishikawa, K.; Utsuka, K.; Tsuchiya, A.; Ando, T. Presented at the *26th International Liquid Chromatography Symposium*, Kyoto, Japan, Jan. 25–26, **1983**.

[21] Reijenga, J.C.; Aben, G.V.A.; Verheggen, T.P.E.; Everaerts, F.M. *J. Chromatogr.* **1983**, *260*, 241.

[22] Tsuda, T. *J. High Resolut. Chromatogr.* **1987**, *10*, 622.

[23] Huang, X.; Luckey, J.A.; Gordon, M.J.; Zare, R.N. *Anal. Chem.* **1989**, *61*, 766.

[24] Issaq, H.J.; Atamna, I.Z.; Muschik, G.M.; Janini, G.M. *Chromatographia* **1991**, *23*, 155.

[25] Issaq, H.J.; Atamna, I.Z.; Metral, C.J.; Muschik, G.M. *J. Liq. Chromatogr.* **1990**, *13*, 1247.

[26] Atamna, I.Z.; Metral, C.J.; Muschik, G.M.; Issaq, H.J. *J. Liq. Chromatogr.* **1990**, *13*, 2517.

[27] Atamna, I.Z.; Metral, C.J.; Muschik, G.M.; Issaq, H.J. *J. Liq. Chromatogr.* **1990**, *13*, 3201.

[28] Tran, A.D.; Park, S.; Lisi, P.; Huynh, O.T.; Ryall, R.R.; Lane, P.A. *J. Chromatogr.* **1991**, *542*, 459.

[29] Green, J.S.; Jorgenson, J.W. *J. Chromatogr.* **1989**, *478*, 63.

[30] Altria, K.D.; Simpson, C.F. *Chromatographia* **1987**, *24*, 527.

[31] Altria, K.D.; Simpson, C.F. *Anal. Proc.* **1988**, *25*, 85.

[32] Nashabeh, W.; El Rassi, Z. *J. Chromatogr.* **1990**, *514*, 57.

[33] Bruin, G.J.M.; Chang, J.P.; Kuhlman, R.H.; Zegers, K.; Kraak, J.C.; Poppe, H. *J. Chromatogr.* **1989**, *471, 429*.

[34] Tsuda, T. *J. Liq. Chromatogr.* **1989**, *12*, 2501.

[35] Grossman, P.D.; Soane, D.S. *Anal. Chem.* **1990**, *62*, 1593.

[36] Rasmussen, H.T.; McNair, H.M. *J. Chromatogr.* **1990**, *516*, 223.

[37] Cao, J.; Cross, R.F. *J. Chromatogr. A* **1995**, *695*, 297.

[38] Cross, R.F.; Cao, J. *J. Chromatogr. A* **1998**, *809*, 159.

[39] Vindevogel, J.; Sandra, P. *J. Chromatogr.* **1991**, *541*, 483.

[40] Chiesa, C.; Horvath, C. *J. Chromatogr.* **1993**, *645*, 337.

[41] Sepaniak, M.J.; Cole, R.O. *Anal. Chem.* **1987**, *59*, 472.

[42] Janini, G.M.; Metral, C.J.; Issaq, H.J.; Muschik, G.M. *J. Chromatogr. A* **1999**, *848*, 417.

[43] Janini, G.M.; Issaq, H.J. in: Guzman, N.A. (Ed.), *Capillary Electrophoresis Technology*, Marcel Dekker, New York, **1993**, 119.

[44] Oda, R.P.; Landers, J.P. in Landers, J.P. (Ed.), *Handbook of Capillary Electrophoresis*, 2nd ed., CRC Press, Boca Raton, **1997**, 28.

[45] Kasicka, B. *Electrophoresis*, **1999**, *20*, 2084.

[46] Wheat, T.E. *Mol. Biotechnol.* **1996**, *5*, 263.

[47] Rickard, E.C.; Towns, J.K. *Methods Enzymol.* **1996**, *271*, 237.

[48] Dolnik, V. *Electrophoresis* **1999**, *20*, 3106.

[49] Wehr, T.; Rodriguez-Diaz, R.; Liu, C.M. *Adv. Chromatogr.* **1997**, *37*, 237.

[50] Janini, G.M.; Fisher, R.J.; Henderson, L.E.; Issaq, H.J. *J. Liq. Chromatogr.* **1995**, *18*, 3617.

[51] Lauer, H.H.; McManigill, D. *Anal. Chem.* **1986**, *58*, 166.

[52] Cohen, N.; Grushka, E. *J. Chromatogr. A* **1994**, *678*, 167.

[53] Bossi, A.; Righetti, P.G. *J. Chromatogr. A* **1999**, *840*, 117.

[54] Corradini, D. *J. Chromatogr. B* **1997**, *699*, 221.

[55] Dolnick, V. *Electrophoresis* **1997**, *18*, *2353*.

[56] Miller, C.; Rivier, J. *J. Pept. Res.* **1998**, *51*, 444.

[57] Deng, B.; LU, J.; Yan, J. *J.Chromatogr. A* **1998**, *803*, 321.

[58] Castagnola, M.; Cassiano, L.; Messana, I.; Paci, M.; Rossetti, D.V.; Giardina, B. *J. Chromatogr. A* **1996**, *735*, 271.

[59] Zhang, R.; Zhang, H.X.; Eaker, D.; Hjerten, S. *J. Capillary. Electrophor.* **1997**, *4*, 105.

[60] Rathore, A.S.; Horvath, C. *J. Chromatogr. A* **1998**, *796*, 367.

[61] Rathore, A.S.; Horvath, C. *Electrophoresis* **1998**, *19*, 2285.

[62] Rathore, A.S.; Horvath, C. *J. Chromatogr. A* **1998**, *796* 367.

[63] Yuan, H.; Janini, G.M.; Issaq, H.J.; Thompson, R.A.; Ellison, D.K. *J. Liq. Chromatogr and rel. technol.* **2000**, *23*, 127.

[64] Baars, M.J.; Patonay, G. *Anal. Chem.* **1999**, *71*, 667.

[65] Matsubara, N.; Terabe, S. *Chromatographia* **1992**, *34*, 493.

[66] Matsubara, N.; Koezuka, K.; Terabe, S. *Electrophoresis* **1995**, *16*, 580.

[67] Koezuka, K.; Ozaki, H.; Matsubara, N.; Terabe, S. *J. Chromatogr.* **1997**, *689*, 3.

[68] Liu, J.; Cobb, K.A.; Novotny, M. *J. Chromatogr.* **1990**, *519*, 189.

[69] Gaus, H.-J.; Beck-Sickinger, A.G.; Bayer, E. *Anal. Chem.* **1993**, *65*, 1399.

[70] Kristensen, H.K.; Hansen, S.H. *J. Liq. Chromatogr.* **1993**, *16*, 2961.

[71] Yashima, T.; Tsuchiya, A.; Morita, O.; Terabe, S. *Anal. Chem.* **1992**, *64*, 2981.

[72] Nashabeh, W.; Greve, K.F.; Kirby, D.; Foret, F.; Karger, B.L.; Relfsnyder, D.H.; Builder, S.E. *Anal. Chem.* **1994**, *66*, 2148.

[73] Strege, M.A.; Lagu, A.L. *Anal. Biochem.* **1993**, *210*, (1993) 402.

[74] Eriksen, J.; Holm, K.A. *J. Capil. Electrophor.* **1996**, *3*, 37.

[75] Tadey, T.; Purdey, W.C. *J. Chromatogr.* **1992**, *583*, 111.

[76] Kats, M.; Richberg, P.C.; Hughes, D.E. *Anal. Chem.* **1995**, *67*, 2943.

[77] Alexander, A.J.; Hughes, D.E. *Anal. Chem.* **1995**, *67*, 3626.

[78] Strickland, M.; Weimer, B.C.; Broadbent, J.R. *J. Chromatogr. A* **1996**, *731*, 305.

[79] James, D.C.; Freedman, R.B.; Hoare, M.; Jenkins, J. *Anal. Biochem.* **1994**, *222*, 315.

[80] M.A. Strege, A.L. Lagu, *Electrophoresis* **1995**, *16*, 642.

Reverse-Phase High Performance Liquid Chromatography and Capillary Zone Electrophoresis Separation of Protegrin Analogs

J. Chen / J. Fausnaugh-Pollitt

Analytical Development Department, IntraBiotics Pharmaceuticals, Inc., 1255 Terra Bella Avenue, Mountain View, CA 94043

3.1 Abstract

Polycationic protegrin analogs were studied by reverse-phase high performance liquid chromatography (RP-HPLC) and capillary zone electrophoresis (CZE). A total of thirty-three peptides were tested to evaluate their separations by HPLC and CZE. The peptides included single amino acid substitutions (D and L isomers), truncated amino- and carboxy-termini, cyclic and all D-amino acid analogs. The separation by CZE was achieved due to charge/mass difference of these peptides. The peptide analog mobility fit the Offord equation. Separation between peptides having the same charge/mass also occurred with some single D-amino acid substituted analogs. A completely different separation profile was obtained by RP-HPLC. RP-HPLC exhibited better separation of the single D-amino acid substituted analogs, while CZE showed better separation of the truncated analogs. The use of HPLC and CZE provides an orthogonal separation tool for protegrin peptides.

3.2 Introduction

Protegrins are a novel class of antimicrobial peptides which have been isolated from porcine leukocytes [1–4] and exhibit broad spectrum antimicrobial activity [5–7]. The protegrins have an amphiphilic, anti-parallel β-sheet structure characterized by two strands bracketing a reverse-turn [8]. One surface of the peptide structure is hydrophobic due to the orientation of amino acid residues such as valine, tyrosine and phenylalanine and the other side is hydrophilic and cationic due to the presence of multiple arginine residues.

RP-HPLC has long been used in the analysis of peptides. The mechanism of separation is through the interaction of hydrophobic amino acids in the peptide and hydrophobic stationary phase groups. It is generally predictive that the more hydrophobic amino acids in the peptide

linear sequence, the longer the RP-HPLC retention time will be [9–11]. This holds true as long as the peptide is short and does not possess significant secondary structure, which could limit the direct interaction of the amino acid residues and the stationary phase.

CZE is a technique that is finding greater utility in the analysis of peptides [12]. The mechanism of separation is based on the peptide's charge/mass ratio. The migration time of related peptides increase with a decrease in charge or an increase in mass.

The differing separation mechanisms of these two techniques offer distinct advantages as orthogonal separations. Idei et al [13] has compared RP-HPLC and Capillary Electrophoresis (CE) in the analysis of somatostatin analogs. They found that CE was comparable and sometimes superior to HPLC. Castagnola et al [14]

compared the CE migration times and RP-HPLC retention times of a tryptic map of skeletal horse myoglobin. They concluded that the two methods do not show a statistical orthogonality with differer selectivity. The comparison of RP-HPLC with CZE work performed by Richards et al [15] on neurohypophyseal peptides and analogs and Sutcliffe et al. [16] on metallothionein isoforms also led to similar conclusions. Gaus et al. [17] used RP-HPLC and CZE or MECC (Micellar Electrokinetic Capillary Electrophoresis) for the separation of basic peptide analogs of neuropeptide Y. RP-HPLC was able to separate the components of some of the complex mixtures, which CZE could not. MECC was needed in those instances to effect a separation by CE. The differences in the peptides of interest need to be evaluated in determining which technique or whether both would have utility.

In this paper, RP-HPLC and CZE have been applied to the separation of a series of protegrin analogs. The techniques were evaluated with respect to their ability to distinguish amino acid substitutions, truncations and modifications of the amino- and carboxy-termini. In some cases, as with the substitution of a D-amino acid, there is no change in the charge/mass ratio, however, there can be a significant change in the three-dimensional structure of the peptide and provide a differing hydrophobic interaction with a RP-HPLC stationary phase. The deletion of an amino acid such as glycine, which is neither significantly hydrophilic nor hydrophobic, can only sometimes be differ-

0009-5893/00/02   27-07 $ 03.00/0

Table I. Peptide Sequences

Peptide Code	Peptide Sequence				
1 (IB-367)	NH₂-	RGG	LCYCRGRFCVCV	GR	CONH₂
2	NH₂-		LCYCRGRFCVCV	GR	CONH₂
3	NH₂-	G	LCYCRGRFCVCV	GR	CONH₂
4	NH₂-	GG	LCYCRGRFCVCV	GR	CONH₂
5	NH₂-	RGG	LCYCRGRFCVCV	GR	COOH
6	Lactyl-	RGG	LCYCRGRFCVCV	GR	CONH₂
7	NH₂-	RGG	LCYCRGRFCVCV	G	COOH
8	NH₂-	GG	LCYCRGRFCVCV	GR	COOH
9	NH₂-	RG	LCYCRGRFCVCV	GR	CONH₂
10	NH₂-	RGG	LCYCRGRFCVCV	R	CONH₂
11	NH₂-	RGG	LCYCRGRFCVCV		COOH
12	NH₂-	RGGR	LCYCRRFCVCV	GR	CONH₂
13	NH₂-	RGGR	LCYCRRFCVCV	GR	COOH
14		(RGGR	LCYCRRFCVCV	GR)	
15	NH₂-		LCYCRRFCVCV	R	COOH
16	NH₂-	RGGG	LCYCRPRFCVCV	GR	COOH
17	NH₂-	RGGR	LCYCRGRFCVCV	GR	CONH₂
18	NH₂-	RGGR	LCYCRGRFCVCV	GR	COOH
19	NH₂-	R	LCYCRPRFCVCV	GR	COOH
20	NH₂-	RGGR	VCYCRGRFCVCV		COOH
21	NH₂-	R	LCYCRRFCVCV		CONH₂
22	NH₂-	R	LCYCRRFCVCV		COOH
23		(R	LCYCRRFCVCV)		
24	NH₂-	**R**GG	LCYCRGRFCVCV	GR	CONH₂
25	NH₂-	RGG	LCYCRGRFCVCV	GR	CONH₂
26	NH₂-	RGG	LCYCRGRFCVCV	GR	CONH₂
27	NH₂-	RGG	LCYC**R**GRFCVCV	GR	CONH₂
28	NH₂-	RGG	LCYC**R**GRFCVCV	GR	CONH₂
29	NH₂-	RGG	LCYCRGRFCVCV	GR	CONH₂
30	NH₂-	RGG	LCYCRGRFCVCV	GR	CONH₂
31	NH₂-	RGG	LCYCRGRFCVCV	GR	CONH₂
32	NH₂-	RGG	LCYCRGRFCVCV	**GR**	CONH₂
33	NH₂-	**RGG**	**LCYCRGRFCVCV**	**GR**	CONH₂

Same disulfide bonds in **1** as in **2–33**. D-Amino acid residues are indicated in **Bold**.

entiated by RP-HPLC, but can be detected in CZE due to the change in mass. The β-sheet structure of the protegrin analogs offer a unique model system for studying the selectivity of these techniques.

3.3 Experimental

3.3.1 Reagents, Apparatus and Methods

Peptide Synthesis

The syntheses of the protegrin analogs (analogs **1–33**) were accomplished following general solid-phase peptide synthesis (SPPS) methodology using either acid labile *t*-Boc (*tert*-butyloxycarbonyl) or base labile Fmoc (9-fluorenylmethoxycarbonyl) as the α-amino protecting groups. All the chemicals and reagents for peptide synthesis using *t*-Boc chemistry were purchased from Bachem, California and the syntheses were carried out on a CS536 Synthesizer (C. S. Bio, San Carlos, CA). All chemicals and reagents for peptide synthesis using Fmoc chemistry were purchased from Perkin-Elmer Applied Bio-

systems and the syntheses were carried out on an ABI 431A Peptide Synthesizer (Applied Biosystems, Foster City, CA). Lactic acid (85% solution in water, A.C.S. reagent grade) for the synthesis of analog **6** was purchased from Aldrich Chemicals.

For *t*-Boc synthesis, the sidechain functional groups of the amino acids were protected as follows: a tosyl group for arginine; a 2-bromobenzyloxycarbonyl group for tyrosine; and a 4-methoxybenzyl group for cysteine. The peptide chain was assembled by coupling the first C-terminal amino acid to 4-methylbenzhydrylamine resin, then removing the *t*-Boc group using trifluoroacetic acid (TFA), neutralizing with triethylamine and coupling the next protected amino acid derivative using dicyclohexylcarbodiimide (DCCI). Repetition of the cycle with the protected amino acids in the desired sequence in reverse order resulted in the fully protected peptide-chain. Couplings were monitored by the ninhydrin test. The crude peptide was obtained by treatment of the peptide-resin with liquid hydrogen fluoride and ethyl methyl sulfide and anisole as scavengers. The crude product was extracted by aqueous acetic acid and cyclized by air

oxidation in ammonium acetate buffer at pH 7.

For Fmoc protection synthesis, the sidechain functional groups of the amino acids were protected as follows: a 2,2, 4,6,7-pentamethyldihydrobenzofuran-5-sulfonyl group for arginine; a *tert*-butyl group for tyrosine; and a trityl group for cysteine. The peptide chain was assembled by coupling the first C-terminal amino acid derivative to 4-(2′,4′-dimethoxyphenyl-Fmoc-aminomethyl)-phenoxy resin, then removing the Fmoc group using piperidine and coupling the next protected amino acid derivative using 2-(1H-benzotriazole-1-yl)-1,1,3,3-tetramethyluronium hexafluorophosphate. Repetition of the cycle with the protected amino acids in the desired sequence in reverse order resulted in the fully-protected peptide-chain. The crude peptide was cleaved from the peptide-resin by TFA with ethanedithiol and anisole as scavengers. The crude product was precipitated by ethyl ether and cyclized by air oxidation in 10% dimethylsulfoxide in Tris buffer at pH 7.

Cyclic analogs were synthesized from peptides that contained a free carboxyl group at the C-terminus. The condensation between the –COOH group and –NH₂ group was accomplished by the coupling agent DCCI in dimethylformide (DMF) solvent at approximately $1–10$ $mg\,mL^{-1}$ concentration overnight (ambient temperature). The coupling was quenched by diluting the reaction mixture with 0.1% TFA. The resultant solution was then subjected to preparative RP-HPLC for purification using the procedure described above. The sequences of the peptides are shown in Table I.

Purification was accomplished by semi-preparative, RP-HPLC on a Vydac C-18 silica column (20 – 50 μm) using gradient elution (22–42% mobile phase B in 70 minutes) with 0.1% TFA as mobile phase A and 0.08% TFA/70% acetonitrile as mobile phase B. The resulting purification fractions with purity greater than 95% as monitored by analytical HPLC were then pooled and lyophilized to yield the TFA salt of the protegrin analogs.

High Performance Liquid Chromatography

Apparatus: HP1100 High performance liquid chromatograph (HPLC) system, Zorbax Stablebond C8 column, 300 Å, 5 μm, 4.6 × 250 mm.

Reagents: acetonitrile, HPLC quality, Burdick & Jackson; de-ionized water,

Milli-Q, polished in-house; TFA, HPLC quality, Pierce Chemicals.

HPLC method: Mobile phase A: 0.15% v/v TFA in water, Mobile phase B: 0.12% TFA/70% acetonitrile (v/v). 0.8 mL min⁻¹ flow rate, column temperature controlled at 20 °C, detection at 214 nm, linear gradient from 30% B to 45% B in 25 min, 38 min run time.

Capillary Zone Electrophoresis

Apparatus: Capillary electrophoresis instrument P/ACE MDQ, Beckman, fused silica capillary, 50 μm I. D., 45 cm effective length (inlet to detector), 55 cm total length. Beckman Φ50 pH/ISE meter.

Reagents: Monobasic sodium phosphate, USP grade, EM Science; de-ionized water, Milli-Q, polished in-house; phosphoric acid, 85% in water, A.C.S. reagent grade, EM Science; membrane filters, 0.22 μm CN (cellulose nitrate) polystyrene. Sodium hydroxide (A.C.S. reagent grade) was purchased from EM Science or used Beckman regenerator solution A (0.1 N NaOH). Hydrochloric acid (A.C.S. reagent grade) was purchased from Aldrich.

CZE Method: Separation at 20 kV for 15 min, buffer rinse before each electrophoresis run, base/acid rinse after each electrophoresis run, detection at 200 nm, capillary temperature controlled at 25 °C, 100 mM sodium phosphate buffer, pH 2.6.

3.3.2 Sample Preparation

IB-367 (analog **1**) was used as the reference peptide throughout the experiments. IB-367 solution was prepared at 0.6 mg mL⁻¹ concentration (based on peptide powder weight) in water, and all analog (analogs **2–33**) stock solutions were prepared at approximate concentrations of 1 mg mL⁻¹ in water. For each individual spiked analog solution, 40 μL of each analog stock solution was added to 1.5 mL of the IB-367 solution. These individually spiked solutions were used to determine the peak assignments in mixtures. For analog mixtures prepared in IB-367 solution, 40 μL of each analog stock solution was added to the IB-367 solution to make a total volume of 1.5 mL. Each analog in the spiked mixtures was present at a 2–4% range relative to IB-367. The peptides present in mixtures 1–5 are as follows:

Table II. Separation of Analogs: Relative Migration Time and Relative Retention Time

Peptide Code	Relative Retention Time	Relative Migration Time	Relative Mobility	Charge (Z) at pH 2.6	Molecular Weight (M)	$Z \cdot M^{\left(-\frac{2}{3}\right)}$
1	1	1	1	5	1900	0.03259
2	1.034	1.057	0.946	4	1629	0.02899
3	1.034	1.073	0.932	4	1686	0.02824
4	1.034	1.087	0.920	4	1743	0.02762
5	1.034	1.057	0.946	4.6	1901	0.03000
6	1.034	1.167	0.857	4	1972	0.02544
7	1.233	1.148	0.871	3.6	1744	0.02485
8	1.070	1.148	0.871	3.6	1744	0.02485
9	1.015	0.987	1.013	5	1843	0.03326
10	0.982	0.990	1.010	5	1843	0.03326
11	1.378	1.147	0.872	3.6	1687	0.02540
12	0.912	0.897	1.115	7	2157	0.04193
13	0.936	0.934	1.071	6.6	2158	0.03952
14	0.959	0.944	1.059	6	2140	0.03613
15	0.967	0.994	1.006	4.6	1673	0.03264
16	0.967	1.123	0.890	4.6	1998	0.02900
17	0.979	0.929	1.076	6	2057	0.03710
18	1	0.985	1.015	5.6	2058	0.03461
19	1	1.039	0.962	4.6	1827	0.03078
20	1.215	1.036	0.965	4.6	1830	0.03075
21	0.879	0.941	1.063	5	1672	0.03549
22	1.317	1	1	4.6	1673	0.03264
23	0.912	1.102	0.907	4	1655	0.02859
24	1	1	1	5	1900	0.03259
25	0.929	1	1	5	1900	0.03259
26	0.648	1.044	0.9579	5	1900	0.03259
27	0.590	1.018	0.9823	5	1900	0.03259
28	0.958	1.018	0.9823	5	1900	0.03259
29	0.692	1.041	0.9606	5	1900	0.03259
30	0.660	1	1	5	1900	0.03259
31	0.660	1	1	5	1900	0.03259
32	0.971	1	1	5	1900	0.03259
33	1	1	1	5	1900	0.03259

Mixture 1: IB-367 with Analog **2, 3, 4, 5, 6, 7, 8, 9, 10,11**

Mixture 2: IB-367 with Analog **12, 13, 15, 16, 17, 18, 19, 20, 22**

Mixture 3: IB-367 with Analog **12, 13, 14**

Mixture 4: IB-367 with Analog **21, 22, 23**

Mixture 5: IB-367 with Analog **24, 25, 26, 27, 28, 29, 30, 31, 32, 33**

3.4 Results and Discussion

Protegrin peptide analogs were studied by RP-HPLC and CZE. The peptide sequences of the tested analogs are shown in Table I. These peptides all have two disulfide bonds that stabilize their secondary structure [18]. Five groups of peptides were studied and their separations are shown in Figures 1–4 and 7. IB-367 was the reference peptide in all the groups. Group 1 (analogs **1–11**) consists of IB-367 analogs where the peptides have –RGRF-β-turn structures and modified N- and C-termini. Group 2 (analogs **12–22**) consists of protegrin analogs that contain different (-turn sequences such as –RRRF-,

–RGRF- and –RPRF-. Group 3 is composed of a set of analogs that have the same peptide sequence but differ in that analog **12** has an amide C-terminus, analog **13** has a carboxyl C-terminus and analog **14** is a cyclic peptide where the N- and C-termini were synthetically joined in an amide bond. Group 4 is another such set of three peptides (analogs **21**, **22** and **23**). Group 5 consists of all the diastereomeric analogs of IB-367, including an all-D analog (analog **33**). The separation results are tabulated in Table II, expressed as the relative HPLC retention time (vs. IB-367) and relative CZE migration time (vs. IB-367).

Separation of Group 1 peptides by CZE and RP-HPLC is shown in Figures 1a and 1b, respectively. All of the analogs are baseline resolved from IB-367 by both techniques. The analogs themselves, however, are not all separated from each other. Those that co-elute in RP-HPLC are resolved by CZE and those that co-elute in CZE are resolved by RP-HPLC. Modifications involving the deletion of glycine residues in the N-terminal end of the molecule are not separated by RP-

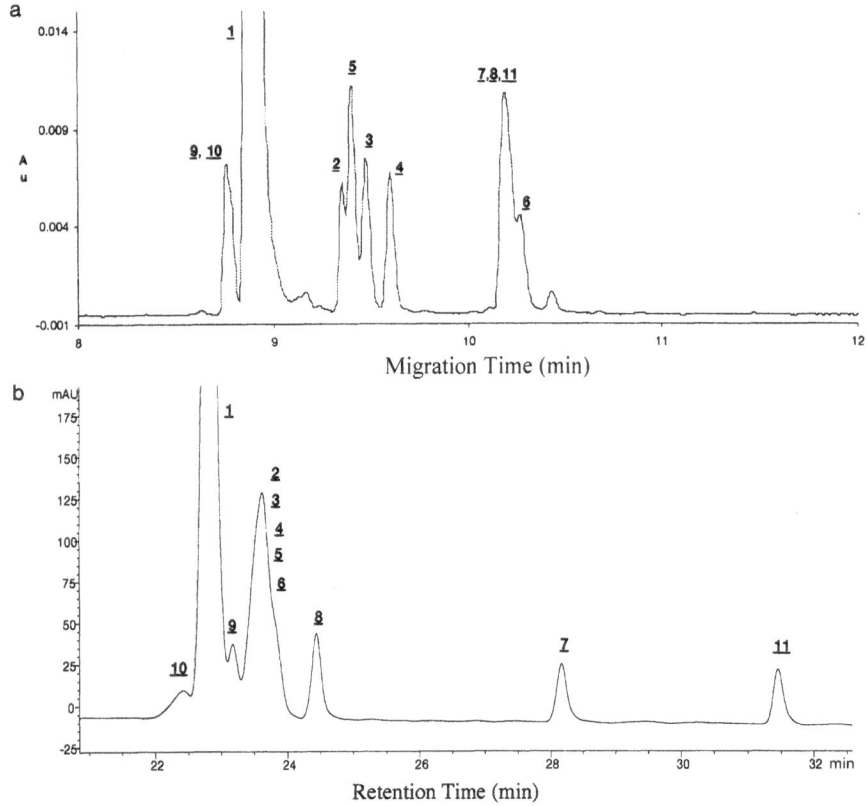

Figure 1. **a** CZE Separation of Analog Group 1. **b** HPLC Separation of Analog Group 1.

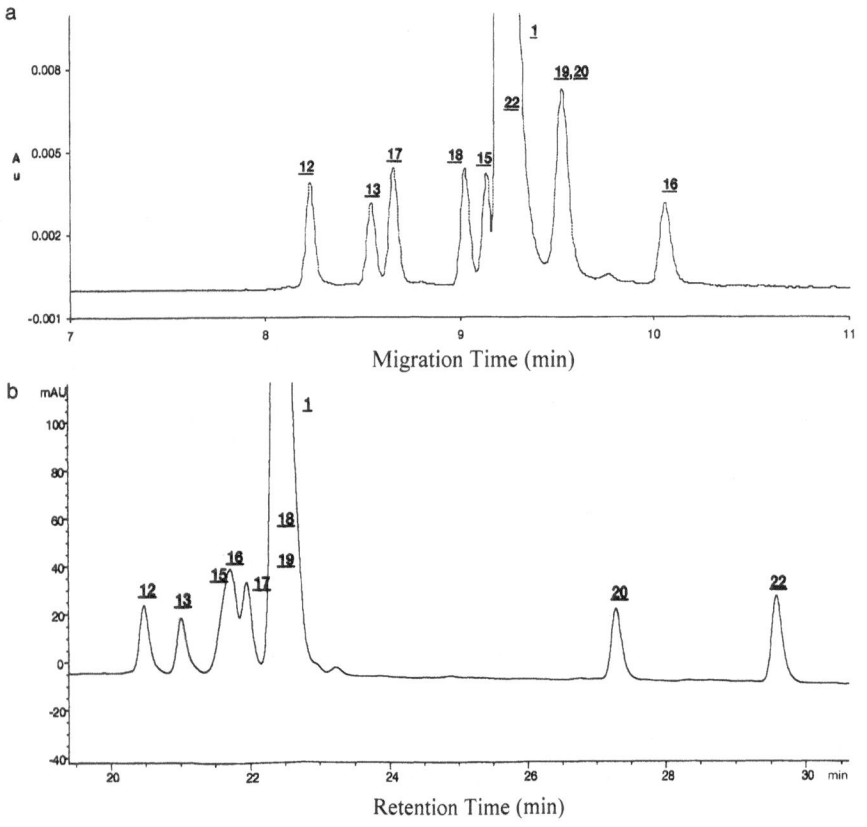

Figure 2. **a** CZE Separation of Analog Group 2. **b** HPLC Separation of Analog Group 2.

HPLC, but are resolved by CZE (analogs 2–4). However, the deletion of a glycine residue in the C-terminal tail of the molecule is separated by RP-HPLC, but not by CZE (analogs 7 and 11). Analogs 9 and 10 could not be resolved by CZE. The difference in these two peptides is the movement of a glycine residue in the sequence from the C-terminal of the molecule to the N-terminal. This change in the peptide sequence appeared to increase the peptide's hydrophilicity as seen by the decreased RP-HPLC retention time.

Separation of Group 2 peptides by CZE and RP-HPLC is shown in Figures 2a and 2b, respectively. Analog pairs 1 and 22 and 19 and 20 could not be separated by CZE. These peptides have very different sequences, but coincidentally have the same charge/mass ratio (Table II). The pairs are resolved from each other by RP-HPLC, however, analogs 1 and 19 co-elute.

The next set of analogs to be evaluated were those in which the C-terminus was either an amide, a carboxylic acid or was cyclized as part of an amide bond (analogs 12, 13 and 14 and 21, 22 and 23). The separation of these peptides by CZE and RP-HPLC is shown in Figures 3 and 4. With group 3 and 4, one can predict that the migration time on CZE would be amide < acid < cyclic, because the charge/mass order is amide > acid > cyclic. In these cyclic peptides, the charge is equal to n, the number of arginine residues in the peptide, since the amine N-terminus is coupled with the carboxyl C-terminus. The prediction is confirmed, as shown in Figures 3a and 4a. The elution order of these peptides by RP-HPLC, however, has no correlation. In group 3, the retention time order was amide < acid < cyclic, while in group 4, the order was amide < cyclic < acid.

Both the RP-HPLC and CZE methods used in this study can resolve deamidation products of the peptides. There were five pairs of acid/amide C-termini peptides analyzed: analogs 1 and 5; 4 and 8; 12 and 13; 17 and 18; 21 and 22. All five pairs of peptides were resolved by RP-HPLC and CZE (Figures 1–4). The separation by CZE was obviously due to their charge difference at pH 2.6. The separation by RP-HPLC indicated the C-terminus was involved in the hydrophobic interactions of the peptides with the HPLC stationary phase.

The orthogonal nature (statistically independent) of these two techniques for the analysis of protegrin analogs can be demonstrated by comparing the RP-HPLC relative retention time and the CZE relative migration time. Figure 5 graphically shows the relationship of these two variables to be a scatter plot with no correlation.

The relative mobility and the calculated charge/mass ratio ($Z \cdot M^{(-2/3)}$) of the protegrin analogs was fit to the Offord equation [19]. A linear plot was obtained with a correlation coefficient (R^2) of 0.97 for the peptides with an amide N-terminus (data not shown). Under the CZE separation conditions, the peptides with an amide N-terminus have a charge of $n + 1$ where n is the number of arginine residues in the peptide and the additional charge is from the N-terminus. For carboxyl C-terminus peptides, the charge at the C-terminus is dependent on the pK_a of the amino acid at the C-terminus. If the pK_a of the individual amino acid [20] is used, the correlation coefficient is only 0.92 for the combined set of peptides tested (amide and carboxyl termini). With the corrections suggested by Rickard et al [21] using a pK_a of 3.2, the correlation coefficient is still only 0.92. A better linear correlation was found ($R^2 = 0.95$) when a pK_a of 2.8 was used for the calculation (Figure 6). This pK_a value resulted in an estimation of a negative charge of 0.4 at the C-terminus of the peptides. Therefore, for the protegrin analogs with a carboxyl C-terminus, the net charge at pH 2.6 is calculated as $n + 1 - 0.4$.

Racemization of amino acids can occur in peptide synthesis as side-reactions [22] and in peptide degradation [23]. Nine peptide analogs of IB-367 having D-amino acid substitutions were synthesized to test the resolving power of these two analytical methods. Most of the single point racemizations in this seventeen amino acid peptide did not result in electrophoretic mobility change, as shown in Figure 7a. This indicated that the substitution did not significantly affect the peptide size. There were two peptides with D-amino acid substitutions that were fully resolved from the IB-367, analog **26**; [D-Tyr[6]] and analog **29**; [D-Phe[11]], and two peptides with D-amino acid substitutions that were partially resolved from IB-367, analog **27**; [D-Arg[8]] and analog **28** [D-Arg[11]]. The sizes of these peptides were, therefore, affected by modifications, which affected the β-sheet structure. In contrast, most of the single D-amino acid substitution peptides were well resolved by RP-HPLC and

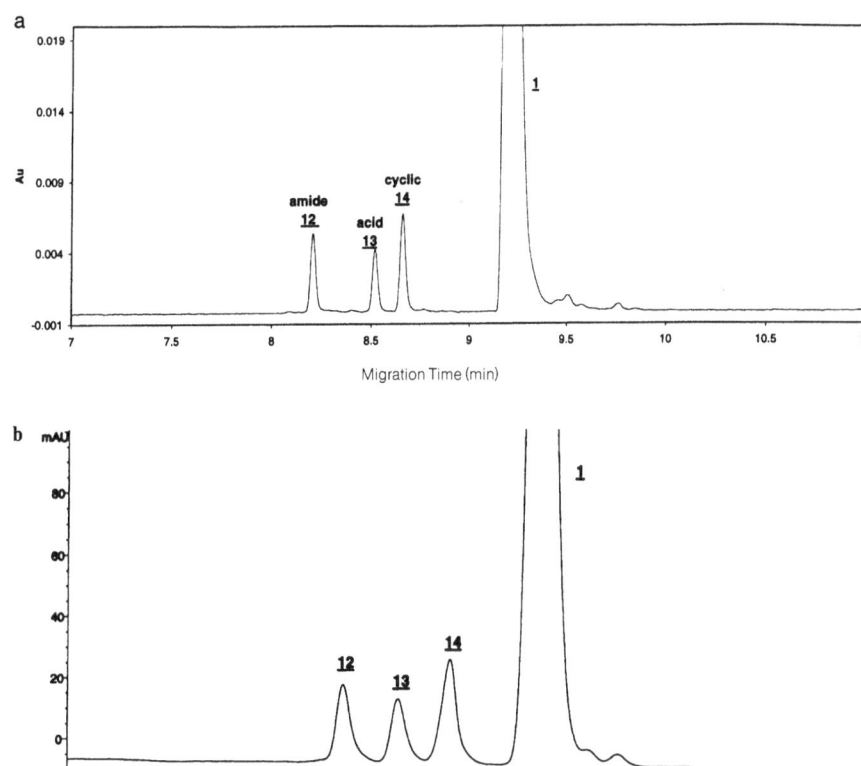

Figure 3. a CZE Separation of Analog Group 3. **b** HPLC Separation of Analog Group 3.

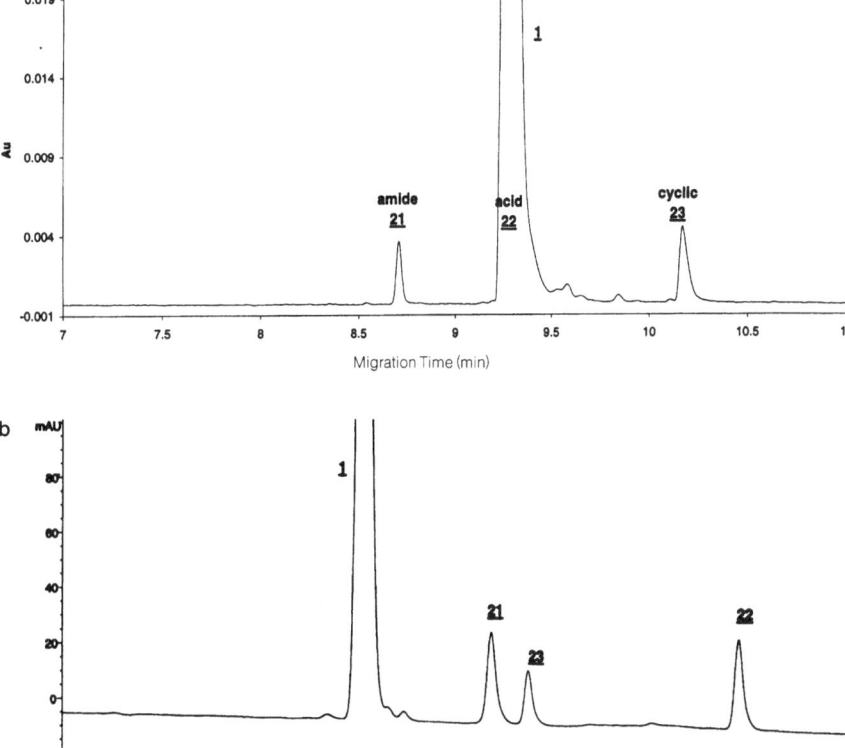

Figure 4. a CZE Separation of Analog Group 4. **b** HPLC Separation of Analog Group 4.

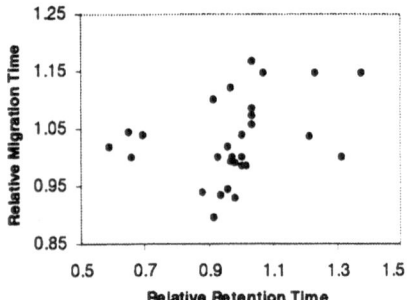

Figure 5. Comparison of HPLC Separation to CZE Separation.

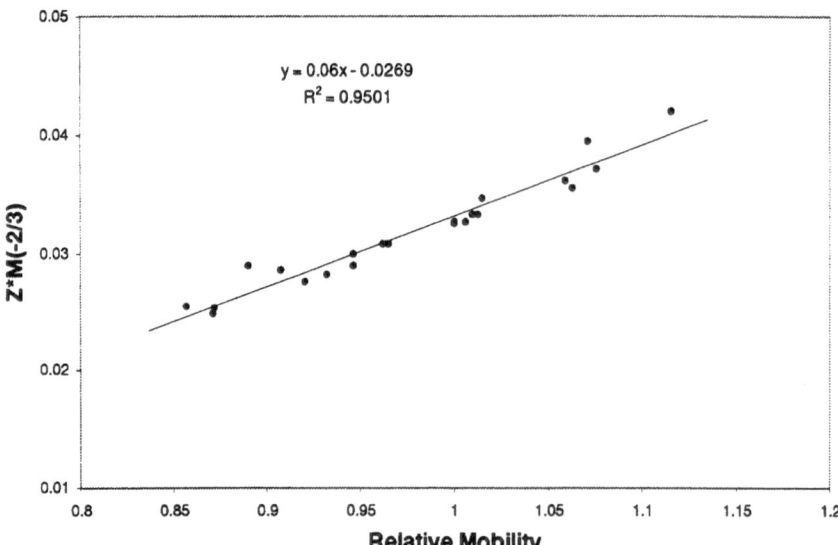

Figure 6. Mobility vs. Charge/Mass Ratio.

exhibited shorter retention times (Figure 7b). The substitutions resulted in decreased hydrophobicity due to the amphiphilic structure of protegrins. The hydrophobic surface of the protegrin molecule interacts with the hydrophobic RP-HPLC stationary phase. Any single D-amino acid substitution would, therefore, result in a decrease in the strength of the hydrophobic interaction of the peptides with the stationary phase. The only exception is analog **24**, where the substitution occurred at the N-terminus residue 1. This again agreed with the previous suggestion that the N-terminus was not as effective as the C-terminus in terms of hydrophobic interactions on RP-HPLC. For analog **32**, where the D-amino acid substitution occurred at the C-terminus, RP-HPLC did show separation.

The separation order of the D-analogs by RP-HPLC suggests that residues Tyr[6] (analog **26**), Arg[8] (analog **27**), Phe[11] (analog **29**), Val[13] (analog **30**), and Val[15] (analog **31**) contribute to the hydrophobic surface of the amphiphilic structure, which is consistent with the solution structure of protegrin [24, 25]. Residue Leu[5] (**25**) is also in the hydrogen-bonding region of the β-sheet, but its effect on the hydrophobic interactions with the RP-HPLC stationary phase is less than that of the other hydrophobic residues. Residues Arg[1] (analog **24**), Arg[10] (analog **28**) and Arg[17] (analog **32**) are not involved in the hydrogen-bonding region of the β-sheet, and therefore, have little effect on the hydrophobic surface of the amphiphilic structure and are eluting close to the parent peptide (IB-367). The all D-analog of IB-367 (analog **33**) has the same hydrophobicity and mass/charge ratio as IB-367, and has no separation from IB-367 by RP-HPLC and CZE.

The β-sheet structure of the protegrin family of peptides and its analogs result in the amphiphilic nature of protegrins: one

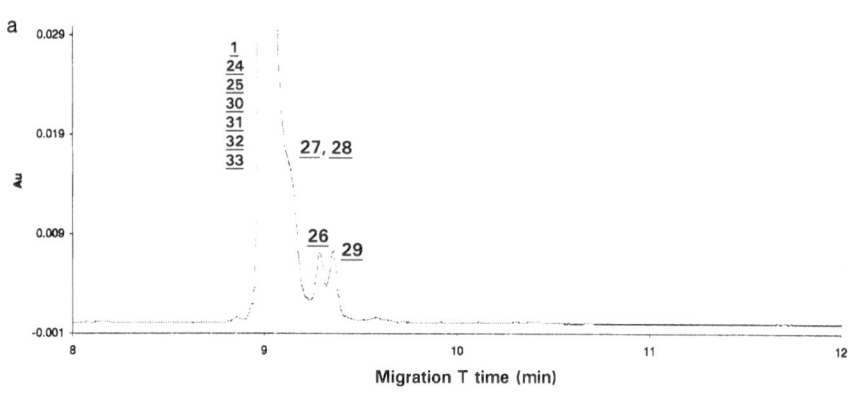

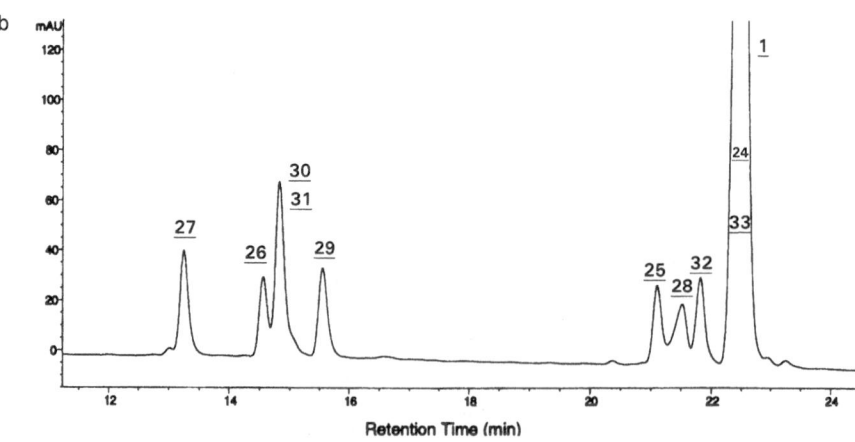

Figure 7. a CZE Separation of D-analog Group 5. **b** HPLC Separation of D-analog Group 5.

surface being net hydrophobic and the other surface being net hydrophilic in character. As demonstrated in the above discussion, this makes it a very good model system for comparing the selectivity of RP-HPLC and CZE. The hydrophobic surface of the protegrin peptides interacts with the hydrophobic stationary phase of the RP-HPLC column. Changes to that hydrophobic surface affect the resolution

by RP-HPLC. The separation mechanism of CZE is different from that of RP-HPLC and peptides resolve from each other due to their difference in charge/size ratio.

3.5 References

[1] Kokryakov, V.N.; Harwig, S.S.L.; Panyutich, E.A.; Shevchenko, A.A.; Aleshina, G.M.; Shamova, O.V.; Korneva, H.A.; Lehrer, R. *FEBS Lett.* **1993**, *327*, 231.

[2] Migorodskaya, O.A.; Shevchenko, A.A.; Abdalla, K.O.; Chernushevich, I.V.; Egorov, T.A.; Musoliamov, A.X.; Kokryakov, V.N.; Shamova, O.V. *FEBS Lett.* **1993**, *330*, 339.

[3] Zhao, C.; Liu, L.; Lehrer, R. *FEBS Lett.* **1994**, *346*, 285.

[4] Zhao, C.; Ganz, T.; Lehrer, R. *FEBS Lett.* **1995**, *368*, 197.

[5] Steinberg et al. "Protegrins: Fast Acting Bacterial Peptides", presented at 8th Intl. Symposium on Staphylococci and Staphylococcal Infection, France, June 1996.

[6] Steinberg et al. "Broad Spectrum Antimicrobial Activity of Protegrin Peptides", presented at 36th International Conference on Antimicrobial Agents and Chemotherapy, New Orleans, La., September **1996**.

[7] Loury, D.J.; Embree, J.R.; Steinberg, D.A.; Stephen, S.T.; Fiddes, J.C. In "Local Application of the Antimicrobial Peptide IB-367 Reduces the Incidence and Severity of Oral Mucositis in Hamsters", Oral Surgery, Oral Pathology, Oral Radiology, and Endodontics, in press.

[8] Harwig, S.; Swiderek, K.; Lee, T.; Lehrer, R. *J. Pept. Sci.* **1996**, *3*, 207.

[9] Meek, J.L. *Proc. Natl. Acad. Sci. U.S.A.* **1980**, *17(3)*, 1632.

[10] Meek, J.L.; Rossetti, Z.L.; *J. Chromatogr.* **1981**, *211*, 15.

[11] Sasagawa, T.; Tsuneo, T.; Teller, D.C. *J. Chromatogr.* **1982**, *240*, 329.

[12] Goor, T.V.-D.; Apffel, A.; Chakel, J.; Hancock, W. in "Handbook of Capillary Electrophoresis", 2nd ed, Edited by James Landers, CRC Press, **1996**, Chapter 8 and references therein.

[13] Idei, M.; Mezo, I.; Vdasz, Z.S.; Horvath, A.; Teplan, I.; Keri, G.Y. *J. Chromatogr.* **1993**, *648*, 251–256.

[14] Castagnola, M.; Cassiano, L.; Rabino, R.; Rossett, D.V.; Andreas, B.F. *J. Chromatogr.* **1991**, *572*, 51.

[15] Richards, M.; Beattie, J. *J. Chromatogr.* **1993**, *648*, 459.

[16] Sutcliffe, N.; Corran, P. *J. Chromatogr.* **1993**, *636*, 95.

[17] Gaus, H.-J.; Beck-Sickinger, A.G.; Bayer, E. *Anal. Chem.* **1993**, *65*, 1399–1405.

[18] Chang, C.; Gu, C.L.; Chen, J.; Steinberg, D.; Lehrer, R.; Harwig, S.S.L. *US Patent*, #5994306, issued 11/30/99.

[19] Offord, R.E. *Nature* **1966**, *591*–*593*.

[20] Dawson, R.M.C.; Elliott, D.C.; Elliott, W.H.; Jones, K.M. *Data for Biochemical Research* (2nd ed.), pp. 1–63, Oxford University Press **1969**.

[21] Rickard, E.C.; Strohl, M.M.; Nielsent, R.G. *Anal. Chem.* **1991**, *197*, 197–207.

[22] Kemp, D.S. in "The Peptides", vol 1 (Gross, E.; Meienhofer, J. Eds.), Academic Press, New York, **1979**, p315–383.

[23] Lauffer, M.A. *Entropy Driven Processes in Biology*, Springer-Verlag, New York, **1975**.

[24] Fahrer, R.L.; Dieckmann, T.; Harwig, S.S.L.; Lehrer, R.I.; Eisenberg, D.; Feigon, J. *Chemistry and Biology*, vol 3, No. 7, p 543–550, **1996**.

[25] Aumelas, A.; Mangoni, M.; Roumest, C.; Chiche, L.; Despaux, E.; Grassy, G.; Calas, B.; Chavanieu, A. *Eur. J. Biochem.* **1996**, *237*, 575–583.

Use of Capillary Electrophoresis to Determine the Dilute Protein Concentration in Formulations Containing Interfering Excipients

S. S. Park[1] / A. Cate / B. S. Chang

Department of Pharmaceutics and Drug Delivery, Amgen Inc., Thousand Oaks, CA 91320, USA

4.1 Abstract

Excipients like human serum albumin (HSA) or surfactants are often added to prevent non-specific adsorption of proteins to surfaces. An enzyme-linked immunosorbent assay (ELISA) has been routinely used to quantify proteins when such excipients interfere with conventional biochemical assays, e. g., UV and HPLC, and make the accurate determination of low protein concentrations and purity difficult. Although the ELISA is a very sensitive assay, the results have large experimental errors contributed by the complicating nature of the assay. In addition, ELISA does not provide information about the qualitative degradation profile of protein, e. g., aggregation, cleavage, and deamidation. As an alternative to the ELISA, a novel capillary electrophoresis (CE) method has been developed to determine both the purity and quantity of Infergen® (Interferon alfacon-1) formulated with interfering excipients like HSA. Results obtained from the CE method were consistent with the results from ELISA, but the CE assay provided more reproducible and precise results. The optimized CE method was successfully applied to the formulation development by determining the recovery of diluted Infergen® in various formulations.

4.2 Introduction

Capillary electrophoresis (CE) has been gradually becoming a powerful technique to solve analytical problems in many biological and chemical disciplines. Although there has been general concern that the CE method is not as reproducible as HPLC for routine analysis, several recent reports [1–6] have shown that CE can be optimized to generate high quality data in terms of precision, accuracy, and linearity. For example, Altria et al. [6] reported a good performance of CE, in terms of precision of peak area and migration time, linearity and accuracy of a chiral analysis of Clenbuterol [7]. These authors also showed repeatability and transfer of CE methods across a variety of applications throughout the pharmaceutical industry [7].

The surface adsorption of proteins is a well known phenomenon [8]. The problem is more significant when the proteins have higher affinity to the surface and/or the protein needs to be diluted to lower concentrations. Infergen® was used as a model protein. This is a very potent protein and is needed to be delivered at very low concentration. Since the amount of Infergen® lost due to surface adsorption is significant at such a low dose; additives like HSA need to be added in the formulation. However, addition of such excipients sometimes renders conventional analyses, e. g., UV, HPLC methods, FTIR, CD,

fluorescence, practically impossible because the concentration of HSA can be 100 fold higher than that of Infergen®.

The separation of protein drug from the HSA can be achieved by using capillary zone electrophoresis based on the difference in charge-to-mass ratio [9]. Coating the capillary with hydrophobic fluorocarbon to form an admicellar bilayer can eliminate the artifact associated with the adsorption of protein drug to the capillary wall [10]. Since coating the capillary reverses the EOF, the voltage applied had to be reversed for the coated capillary in order for all of the components to migrate out of the capillary [10]. For the separation of protein drugs, it is desirable to have a relatively long capillary and a highly concentrated buffer solution so that the resolution and separation of the sample is maximized [11]. The long capillary allows for higher voltages to be applied without having to worry about the effects of Joule heating on the analyte. The higher voltage in turn increases the resolution by shortening the run time thus minimizing band broadening caused by diffusion. The high buffer concentration increases the efficiency by contributing to the focusing effect [10]. With these optimization parameters in mind, the separation and quantification of protein drugs can be easily attained.

In this study, a novel CE method was developed to determine low concentrations ($10-30 \, \mu g \, mL^{-1}$ range) of a protein drug formulated with interfering excipient like HSA. This method can separate the protein from the interfering excipients and allow accurate determination of pro-

[1] To whom correspondence should be addressed at: One Amgen Center Drive, Mailstop 8-1-C, Thousand Oaks, CA 91320, USA

0009-5893/00/02   34-05 $ 03.00/0

tein concentration. The results obtained with the CE method were compared with results from enzyme-linked immunosorbent assay (ELISA) to demonstrate the validity of the CE method. In addition, the reproducibility of the standard curve, peak area, and migration time obtained by the CE method was observed over several days to confirm that the method can be used as a routine analytical method. The CE method was further applied to the formulation development by monitoring the recovery and purity of the protein drug after exposing it to a brief stress.

Infergen®, a model protein used in this study, is a recombinant, non-naturally occurring type-I interferon. Scanning the sequences of several natural interferon alpha subtypes and assigning the most frequently observed amino acid in each corresponding position has derived the 166-amino acid sequence of Interferon alfacon-1. Four additional amino acid changes were made to facilitate the molecular construction [12, 13]. Infergen® is approved for the treatment of chronic hepatitis C virus (HCV) infection in patients 18 years of age or older with compensated liver disease who have anti-HCV serum antibodies and/or the presence of HCV RNA. Infergen® is presented as a $30 \, \mu g \, mL^{-1}$ concentration to accommodate the therapeutic doses of 9 and 12 μg.

4.3 Material and Methods

Infergen® was manufactured and purified at Amgen (Thousand Oaks, CA, USA). Human Serum Albumin was purchased from Dept of Health (Michigan Dept. of Health, USA). Bovine serum albumin was purchased from Sigma Chemical Company (St. Louis, MO, USA). Polybrene used for capillary coating was obtained from Sigma Chemical Co. (St. Louis, MO, USA). Other chemicals used for the study were analytical grade or better.

CE analysis was performed with a Crystal CE 310 (Thermo CE, Franklin, MA, USA) capillary electrophoresis system equipped with a Class-VP chromatography data system (Shimadzu Scientific Instruments, Columbia, MD, USA) for data analysis. A Perkin Elmer 785A UV/VIS detector (Norwalk, Conn., USA) was employed for monitoring all separations performed in this study. A fused, uncoated silica capillary of 50 μm id and 95 cm length (75 cm effective length from the injector to detector) was used. The ap-

plied voltage was $-15 \, kV$ (reversed charged). The oven temperature was 25 °C and sample tray temperature was 4 °C. The injection of sample was done at 50 mbar for 0.2 min (9.7 nL). The capillary was first washed with 0.1 N hydrochloric acid, water, 1 N sodium hydroxide, followed by 0.1 N sodium hydroxide before coating with 5% polybrene (Sigma, St. Louis, MO, USA) mixture. The wash and coating cycle was performed before each CE run. A solution of 150 mM sodium phosphate buffer was used for the separation of protein in the formulation.

4.3.1 Quantification of Infergen® in Formulation Samples

The Infergen® bulk standard concentrations from $2-200 \, \mu g \, mL^{-1}$ in 25 mM sodium phosphate, 100 mM sodium chloride at pH 7 buffer were analyzed by CE. The total area for the three isoforms peak for each concentration was plotted against the Infergen® concentration to prepare the standard curve. Infergen® concentrations for samples were determined by interpolation on this standard curve. Stability study was carried out by monitoring the recovery of Infergen® in a Mini-Med external pump catheter during storage at 37 °C. The Infergen® concentration determined before introduction to the catheter was used as a control.

4.4 Results and Discussion

4.4.1 Separation of Infergen® from Formulation Ingredients

The critical advantage of using CE is its high resolving power, which allows the separation of heterogeneous species of Infergen® from other interfering ingredients in the formulation. A major challenge is to detect small amounts of Infergen® in the presence of large amount of albumin, which are at least two orders of magnitude higher in terms of concentration. With the optimized CE method, Infergen® could be separated well from the albumin and several isomeric forms of Infergen® could be detected (Figure 1). The result demonstrates that CE can not only be used for detecting native Infergen® species but also for distinguishing minor variations in the protein. The slight change in the migration time of HSA or Infergen® when they were mixed, is a typical phenomenon in

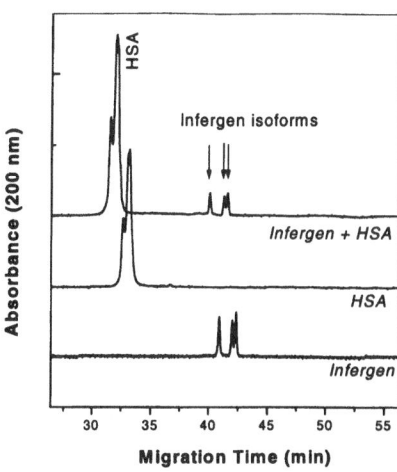

Figure 1. Capillary electropherograms showing the separation of Infergen® isomers from human serum albumin. Infergen® concentration: $100 \, \mu g \, mL^{-1}$ in 0.25% HSA, Injection volume: 9.7 nL. Running buffer: 150 mM sodium phosphate pH 6.0.

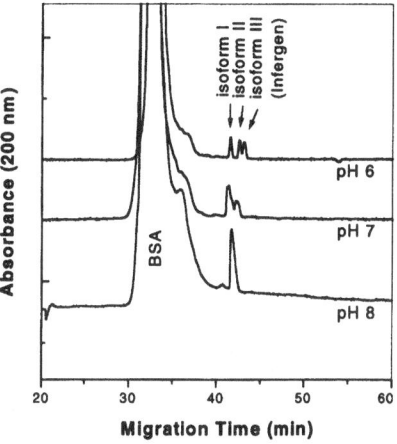

Figure 2. Effect of the pH of running buffer on the resolution of CE analysis. Phosphate buffer at 150 mM was used to control the pH. Same samples were performed at the three pHs.

CE where every component in a sample contributes to the migration of protein molecules, as the total conductivity of the sample affects the overall field strength experienced across the capillary. Similar changes in migration time can be the result of minor changes in the composition of a protein formulation.

4.4.2 Selection of pH for Running Buffer

To optimize the CE condition, the effect of the pH of running buffer on the resolution of Infergen® was tested. Figure 2 shows that Infergen® was well separated from BSA at all tested pHs of 6–8, but different Infergen® isoforms were better resolved at pH 6.0. Since the CE assay will

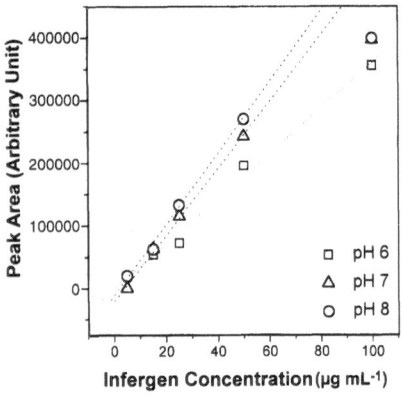

Figure 3. Effect of the pH of running buffer on the linearity between Infergen[R] concentration and total peak area.

be used to determine Infergen[®] concentration, the linearity of peak area (total area of all isoforms) obtained from different pHs was also examined. As shown in Figure 3, linear correlation between the peak area of Infergen[®] and its concentration was observed up to 100 µg mL^{-1} of Infergen[R] when pH 6 was used as the running buffer. When pH 7 or pH 8 was used as the running buffer, a relatively larger peak area was observed at lower Infergen[®] concentrations but the linearity was only maintained up to 50 µg mL^{-1} of Infergen[®] (Figure 3). Since the determination of protein concentration was one of the

main objectives of this study, pH 6 buffer was selected as the running buffer.

4.4.3 Reproducibility of Migration Time

In order to determine the precision of migration time, migration time of each isoform was compared for inter-day variation and intra-day variation. A total of five analyses were carried out per day for 6 different days. Table I shows the inter- and intra-day reproducibility of migration time. Within a day (intra-day), the relative standard deviation (RSD) for the precision of migration time was between 0.3% and 1.8%. However, the relative standard deviation (RSD) for the precision of inter-day migration time was between 2.6% and 3.6%. It must be noted that the migration rate of proteins during capillary electrophoresis is extremely sensitive to a minor variation in sample composition or running buffer.

4.4.4 Reproducibility of Peak Area

Within a day (intra-day), the relative standard deviation (RSD for $n = 5$) for the precision of peak area of Infergen[®] was between 2.9 and 6.5%. However, the relative standard deviation (RSD for $n = 6$) for the inter-day precision of peak area was 7.6% (Table II). The results suggest that the CE method will have less experimental error that ELISA method which has generally 20% RSD.

4.4.5 Reproducibility of Calibration Curve

The reproducibility of the linear regression coefficient of peak area over the Infergen[®] concentration range between 1 µg mL^{-1} and 200 µg mL^{-1} was examined.

Results obtained over 10 days showed that the relative standard deviations (RSD) of regression coefficient was 9.5% (Table III). The r value of the curve fitting turned out to be higher than 0.994 with RSD of 0.3%. The results indicate that the protein concentration determined by the CE method will be more accurate than routine biological assay methods like ELISA.

Table I. Inter- and Intra-day precision of migration time (MT) for Infergen[®] isomers.

A) Intra-day precision					
Day	Isoforms	Runs (n)	Average MT (min)	Std Dev	RSD (%)
1	Isoform I	5	42.52	0.13	0.301
1	Isoform II	5	43.82	0.13	0.29
1	Isoform III	5	44.39	0.14	0.31
2	Isoform I	5	43.81	0.11	0.25
2	Isoform II	5	45.21	0.11	0.25
2	Isoform III	5	45.82	0.12	0.26
3	Isoform I	5	40.10	0.66	1.64
3	Isoform II	5	41.24	0.70	1.71
3	Isoform III	5	41.71	0.74	1.78
4	Isoform I	5	40.61	0.54	1.32
4	Isoform II	5	41.81	0.57	1.36
4	Isoform III	5	42.30	0.58	1.38
5	Isoform I	5	41.27	0.29	0.71
5	Isoform II	5	42.48	0.34	0.80
5	Isoform III	5	43.00	0.34	0.79
6	Isoform I	5	40.42	0.22	0.53
6	Isoform II	5	41.73	0.29	0.70
6	Isoform III	5	42.23	0.31	0.74

B) Inter-day precision				
Isoforms	Days (n)	Average MT (min)	Std Dev	RSD (%)
Isoform I	6	41.45	1.44	3.47
Isoform II	6	42.48	1.11	2.60
Isoform III	6	43.24	1.57	3.63

Table II. Inter- and Intra-day precision of total peak area for Infergen[®].

A) Intra-day precision				
	Runs (n)	Average Peak Area	Std Dev	RSD(%)
Day 1	5	41060	1167	2.84
Day 2	5	43248	2139	4.95
Day 3	5	47742	3122	6.54
Day 4	5	49179	1807	3.67
Day 5	5	47964	1856	3.87
Day 6	5	49767	2172	4.36

B) Inter-day precision			
Runs (n)	Average Peak Area	Std Dev	RSD(%)
6	46493	3513	7.56

Table III. Reproducibility of CE data of Infergen® standard calibration curve over 10 days.

Day	Slope	r
1	5082.6	0.994
2	4959.3	0.995
3	5072.6	0.998
4	4133.4	1.000
5	3994.1	0.995
6	4781.4	0.998
7	4962.8	0.995
8	3999.0	0.999
9	4905.0	0.996
10	4658.6	0.990
Average	4654.9	0.996
Std dev	442.5	0.003
RSD	9.51%	0.30%

Table IV. Effect of HSA concentration on the recovery of Infergen determined by CE and ELISA after incubation in a Mini-Med external pump catheter. Infergen® concentration was 20 µg mL^{-1}.

HSA Concentration (%)	CE		ELISA	
	0 Day*	1 Day (37 °C)	0 Day*	1 Day (37 °C)
0	56%	29%	27%	44%
0.1	85%	54%	78%	47%
0.25	95%	48%	89%	54%
0.5	92%	52%	92%	57%
1.0	92%	55%	107%	74%

* The data were obtained from the sample taken immediately after introduction into the catheter for day 0. Recovery was based on the control sample that is obtained before introduction to the catheter. Each value is the mean of three replicates.

Table V. RSD comparison between CE and ELISA data of Infergen® showed in Table IV.

HSA Concentration (%)	CE		ELISA	
	0 Day	1 Day (37 °C)	0 Day	1 Day (37 °C)
0.00	8.3%	2.1%	52.7%	33.1%
0.10	8.5%	6.5%	15.7%	9.7%
0.25	6.4%	5.3%	0.9%	4.5%
0.50	9.0%	4.9%	7.2%	74.1%
1.00	7.4%	4.0%	3.7%	7.2%

RSD is based on three replicate determinations.

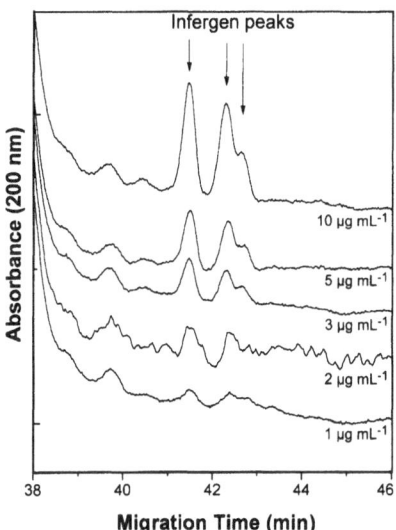

Figure 4. Capillary Electropherograms obtained with samples containing various Infergen® concentrations. Running buffer: 150 mM sodium phosphate, pH 6.0.

4.4.6 Detection Limit

Figure 4 shows the electropherograms of Infergen® injected at various concentrations. With the injection volume of 9.7 nL as discussed in the Method section, the lowest detection level of the sample in 0.5% BSA solution was 1 µg mL^{-1}. The sensitivity is sufficient to analyze the formulations containing 10–30 µg mL^{-1} of Infergen®.

4.4.7 Application of the CE Method to Actual Formulations

Effect of Human Albumin on the Surface Adsorption of Infergen®

In this study, the formulation appropriate for the continuous infusion of Infergen® by using a Mini-Med external pump de-

vice was developed. Infergen® is a highly potent drug and is formulated at low concentrations around 30 µg mL^{-1}. In addition, Infergen® appears to be susceptible to nonspecific surface adsorption like other proteins. Since the amount of surface adsorption loss is significant at such a low protein concentration of Infergen®, the formulation containing human albumin (HSA) as a stabilizer was considered. To study the efficacy of albumin in terms of preventing Infergen® from sticking to the surface of a Mini-Med external pump device, Infergen® at 20 µg mL^{-1} was formulated with various concentrations of HSA. Samples were analyzed by both CE and ELISA immediately after filling into the Mini-Med external pump catheter and after incubation for 1 day at 37 °C. The results are shown in Table IV. Analyses immediately after filling indicate that HSA effectively inhibits Infergen® from surface adsorption as indicated by higher Infergen® concentration in HSA-containing formulations. However, samples stored at 37 °C showed approximately 50% loss regardless of the HSA concentration, suggesting that a different mechanism is involved in the recovery loss when samples are exposed to thermal stress. It is possible that the changes in physical properties of Infergen® associated with structural changes at higher temperature can contribute to the additional adsorption to the surface. The results obtained by CE were consistent with ELISA results considering the variability of both assays. Percent CV

data indicate that the CE data was more reliable than the ELISA data (Table V). Some of the ELISA data showed 30–75% CV between the runs, indicating significant variability of the assay. On the other hand, CE results turned out to be less variable with less than 10% CV value (Table V).

In order to understand the effect of Infergen® on the surface adsorption in the presence of HSA, various concentrations of Infergen® (10–200 µg mL^{-1}) formulated with HSA (0.1%) were prepared. Samples were analyzed by both CE and ELISA after filling into the syringe and after incubation for 1 day at 29 °C. Both sets of data showed that HSA at 0.1% effectively inhibits Infergen® from sticking to the syringes when Infergen® concentration is higher than 30 µg mL^{-1} (Table VI). CE results were consistent with ELISA data.

Formulation Optimization for Minimal Adsorption Loss

Because HSA was not sufficient to maintain the recovery of Infergen® in the syringe, i.e., only 50% recovery after 1 day at 37 °C even with 1% HSA (data not shown), the investigation of other formulations was necessary to improve recovery of Infergen®. Formulations containing different combinations of polysorbate 80, HSA, sodium chloride, and sorbitol were prepared to find a condition where the sticking of Infergen® in a delivery device

Table VI. Effect of Infergen® concentration on the recovery of Infergen® determined by CE and ELISA after incubation in a Mini-Med external pump catheter in the presence of 0.1% HSA:% recovery against control sample*.

Infergen® Concentration (μg mL⁻¹)	CE		ELISA	
	0 Day	1 Day (29 °C)	0 Day	1 Day (29 °C)
10	101%	51%	96%	54%
30	124%	103%	107%	88%
50	132%	86%	170%	95%
100	88%	84%	101%	80%
200	97%	76%	88%	82%

* Control sample: Before introduction to catheter.

Table VII. Effect of different additives on the recovery of Infergen® in a delivery device. Each formulation includes 20 mM sodium phosphate buffer at pH 7:% recovery against the control* for each of the formulation before contact with the Mini-Med external pump catheter.

Additives	CE		ELISA	
	0 Day	1 Day (37 °C)	0 Day	1 Day (37 °C)
No	100%	45%	102%	48%
100 mM NaCl	99%	52%	120%	41%
30% Sorbitol	110%	27%	87%	33%
0.1% HSA	85%	63%	98%	95%
0.1% HSA, 30% Sorbitol	92%	74%	112%	88%
0.01% polysorbate 80, 30% Sorbitol	104%	106%	102%	102%
0.1% HSA, 0.01% polysorbate80, 30% Sorbitol	96%	88%	99%	95%
0.1% HSA, 0.01% polysorbate 80	99%	109%	110%	102%
0.01% polysorbate 80	106%	103%	108%	101%

* Control samples for each of the formulations were obtained before contact with the Mini-Med external pump catheter. The data for 0 day is obtained from plastic syringe device. The data for 1 day at 37 °C is obtained from the catheter to compare the stability of various formulations.

is minimum. Infergen® was well resolved from all the tested excipients by CE and the determination of Infergen® concentration based on peak area was relatively straightforward. Sorbitol, which is known as a good thermal stabilizer, facilitated the surface adsorption of Infergen® when samples were stored at 37 °C (Table VII). The result with sorbitol implies either that thermal stability does not contribute to the surface adsorption of Infergen® or that there is another dominating factor induced by sorbitol to facilitate the surface adsorption of the protein. All formulations containing polysorbate 80 showed perfect recovery in the delivery device even after incubation for 1 day at 37 °C.

Once again, the results obtained with CE were consistent with the ELISA results. Based on the results, polysorbate 80 was selected as a best stabilizer to prevent Infergen® from surface adsorption loss.

4.5 Conclusion

An analytical method to determine the concentration of diluted Infergen® concentration in various formulations containing interfering excipients was developed with Capillary Electrophoresis. The CE method turned out to be easier to use with the advantages of better accuracy and reproducibility as compared to conventional ELISA method. The novel CE method was successfully applied to the formulation development of Infergen® where decrease in protein concentration due to surface adsorption was a significant problem.

4.6 Acknowledgements

The author wishes to acknowledge Timothy Blanc for helpful discussion and information about the CE method and Concesa Jornacion for ELISA analysis.

4.7 References

[1] Ackermans, M.T.; Beckers, J.L.; Everaerts, F.M.; Seelen, I.G.J.A. *J. Chromatogr.* **1992**, *590*, 341.

[2] Tsai, E.W.; Singh, M.M.; Lu, H.H.; Ip, D.P.; Brooks, M.A. *J. Chromatogr.* **1992**, *626*, 24.

[3] Weinberger, R.; Albin, M. *J. Liq. Chromatogr.* **1991**, *14*, 953.

[4] Swartz, M. *J. Liq. Chromatogr.* **1991**, *14*, 923.

[5] Altria, K.D. *J. Chromatogr.* **1993**, *634*, 323.

[6] Pluym, A.; Van Ael W.; Smet, M. *Trends Anal. Chem.* **1992**, *11*, 27.

[7] Altria, K.D.; Harden, R.C.; Hart, M.; Hevizi, J.; Hailey, P.A.; Makwana J.V.; Portsmouth, M.J. *J. Chromatogr.* **1993**, *641*, 147.

[8] Sadana, A. *Chem. Rev.* **1992**, *92*, 1799.

[9] Rickard, E.C.; Strohl, M.M.; Nielsen, R.G. *Anal. Biochem.* **1991**, *197*, 197.

[10] Emmer, A.; Jansson, M.; Roeraade, J. *J. Chromatogr.* **1991**, *547*, 544.

[11] McLaughlin, *Beckman Technical Information Bulletin.* TIBC-106 **1991**.

[12] Alton, K.; Stabinsky, Y.; Richards, R. *Production, characterization and biological effects of recombinant DNA derived human IFN-α and IFN-γ analogs*, In: De Mayeyer, E.; Schellekens, H. eds. *The Biology of the Interferon System 1983*. Elsevier Science Publishers: Amsterdam. 119–128 **1983**.

[13] Blatt, LM.; Davis, J.; Klein, SB.; Taylor MW. The biologic activity and molecular characterization of a novel synthetic interferon-alpha species, consensus interferon. *J. Interferon Cytokine Res.* **1996**, *16*, 489–499.

Effects of the Solution Environment on the Resolution of Recombinant Human Deoxyribonuclease Variants in Capillary Zone Electrophoresis

C. P. Quan[1] / E. Canova-Davis[1] / A. B. Chen[2]

Departments of Analytical Chemistry[1] and Quality Control[2], Genentech, Inc., One DNA Way, South San Francisco, CA 94080, USA

5.1 Abstract

Capillary zone electrophoresis (CZE) is used to demonstrate the interaction of a protein analyte and the surrounding free solution environment, specifically the effects of hydrogen ion or divalent metal cation titration on the resolution of recombinant human deoxyribonuclease I (rhDNase) variants. Hydrogen ion titration of the CZE buffer solution environment observably alters the surface charges of the protein, leading to measurable changes in the electrophoretic mobility which correlates to the theoretical net average protein charge (Z). Similarly, divalent metal cations, when added to the CZE buffer solution, associate with the acidic rhDNase surface charges and lead to observable changes in electrophoretic mobility. Typically, conditions that led to decreased electrophoretic mobility of the protein also led to enhanced resolution of broad zones associated with the heterogeneity of the protein. Additional characterization of the observed heterogeneity distinguished between a distribution of associated ions and glycosylation variation as the source of the protein-associated heterogeneity.

5.2 Introduction

Recombinant human deoxyribonuclease I (rhDNase) is manufactured as Pulmozyme™, a therapeutic enzyme treatment for human patients with cystic fibrosis [1]. Expressed in Chinese hamster ovary cells as a single chain of 260 amino acids, rhDNase is glycosylated at both asparagine consensus sites with complex, hybrid, and high mannose structures that possess varying amounts of sialylation (0–4 residues at each site) and phosphorylation (0–2 residues at each site) [2]. rhDNase is an acidic protein, with a predicted isoelectric point (pI) of 4.6; the addition of sialylated and/or phosphorylated glycosylation structures generate acidic heterogeneity that appears on an isoelectric focusing (IEF) gel as a ladder of 12–16 bands located between the 3.5- and 4.3-pI markers

[2]. Post-translational glycosylation also increases the apparent size of the molecule, from the calculated molecular weight of 29,254 daltons to almost 42,000 daltons measured by size-exclusion chromatography [2].

Capillary zone electrophoresis (CZE) is a high performance technique that separates analytes based on their surface-charge and hydrodynamic size differences, as these analyte molecules possess different free solution mobilities in an electric field. Versatile and robust, CZE is one of the simplest analytical techniques to perform, requiring only a capillary filled with buffer as the separation medium. However, as has been previously demonstrated, the CZE profile of rhDNase is strongly influenced by both the buffer pH and the addition of calcium counter-ions to the buffer [3]. Furthermore, the ex-

tent of calcium interaction with rhDNase in the CZE mode was shown to be pH dependent, since the electrophoretic mobility of the protein changed with calcium addition to the buffer at basic pH, compared to little change in mobility after the initial addition of calcium to the buffer at acidic pH [3].

The work presented in this paper expands on the observed differences [3] in the electrophoretic mobility of the protein with changes in the buffer solution pH, and directly relates mobility changes to the theoretical charge characteristics of the protein molecule. This work further explores the effect of the addition of various metal ions to the buffer solution that are known to associate with rhDNase under physiological conditions, and that can have structural and/or catalytic roles [4]. The resulting changes to the protein electrophoretic mobility in buffer solutions of different metal ion content are related to the binding affinities of these essential metal ion interactions. Ionic manipulations of the buffer solution environment reveal charge and/or size variations in this glycoprotein profile, which are further analyzed by removal of specific glycosylation-based charge residues and compared to complete deglycosylation.

5.3 Materials

The following chemical reagents were purchased from Sigma-Aldrich (St. Louis, MO, USA): Tris (hydrochloride and base), calcium chloride ($CaCl_2$), sodium chloride (NaCl), magnesium chloride ($MgCl_2$), and

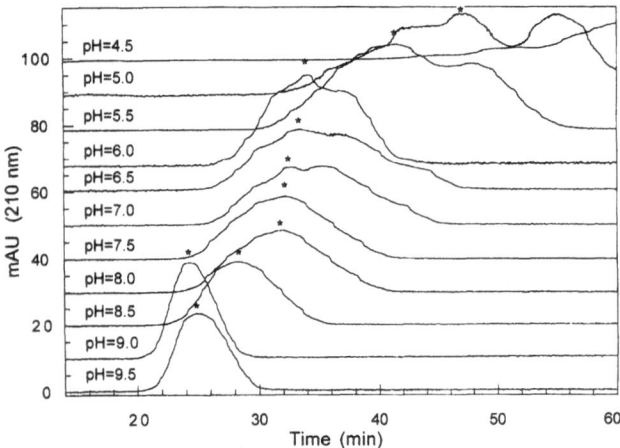

Figure 1. CZE of rhDNase with buffers at different pH. Sodium phosphate buffers (20 mM) were used for pH 9.5 to 6.5 and sodium citrate buffers (20 mM) were used for pH 6.0 to 4.5. Run times were extended to 100 minutes for the pH 5.0 buffer and to 180 minutes for the pH 4.5 buffer. Asterisks indicate the migration times used to plot the experimental pH effects shown in Figure 2.

Table I. Charged amino acid composition of rhDNase.

Charged Residue	Expected pKa Value in Protein [5, 6]	# Residues in rhDNase
Aspartic acid	4.4–4.6	22
Glutamic acid	4.4–4.6	12
Histidine	6.0	5
Lysine	10.4	6
Arginine	12.0	13
N-terminal (leucine)	8.2	1
C-terminal (lysine)	2.2	1

manganese chloride (MnCl₂). Sodium citrate and sodium phosphate buffer solutions (20 mM) were purchased from Fluka BioChemika (Buch, Switzerland). Water purchased from Hewlett-Packard (Palo Alto, CA, USA) was used to prepare all buffers and for sample desalting. Protein modification enzymes were purchased as follows: neuraminidase and alkaline phosphatase from Boehringher Mannheim (Germany), and PNGase F from New England BioLabs (Beverly, MA, USA). The rhDNase protein, prepared at Genentech, was converted to the stable, fully deamidated form at the asparagine-74 site to remove amino acid-based charge heterogeneity.

5.4 Methods

5.4.1 Instrumentation

A Hewlett-Packard (Palo Alto, CA) HP3DCE system, with HP Chemstation software version A.04.01, was used for all experiments. Poly (vinyl alcohol) (PVA) coated capillaries, 50 μm × 64.5 cm (inner diameter by total length, with an effective length of 56 cm), were purchased from Hewlett-Packard.

5.4.2 Charge-Specific Digests

Procedures for the dual, charge-specific digest (neuraminidase plus alkaline phosphatase) have been previously published [3]. Dephosphorylation was determined to be 100% by HPLC analysis; desialylation was assayed at 90%.

5.4.3 Asparagine-Linked Deglycosylation

PNGase F (385 milliunits in 50 μL) was added to 100 μg of rhDNase at 0.5 mg/mL in pH 8.3 digest buffer (10 mM Tris-HCl, 1 mM CaCl₂), and incubated for 24 hours at 37 °C. The digested sample was desalted by dialysis into 2 mM CaCl₂ using Pierce (Rockford, IL) Slide-A-Lyzer cassettes. Complete deglycosylation was verified by analysis of the tryptic digest peptide map by electrospray mass spectrometry, which revealed the absence of the two previously characterized glycopeptide peaks and the appearance of two new peaks with the appropriate masses [2].

5.4.4 Capillary Zone Electrophoresis

A PVA-coated capillary was used to eliminate the effects of electroosmotic flow and to minimize protein interaction with the charged inner wall of the capillary. The capillary was flushed with buffer for 10 minutes at 2 bar, samples were hydrodynamically injected for 20 seconds with a pressure of 50 mbar at the inlet, and then 30 kV was applied in the reverse-polarity mode. The protein sample, at 0.5 mg/mL, was dialyzed into 2 mM CaCl₂ to maintain structural stability for the duration of the experiments. The capillary was maintained at 20 °C, and the absorbancy was monitored at 214 nm. The current was monitored and was consistent for all experiments (approximately 12–20 μA, depending on buffer composition). Information on specific buffers is included with the figure captions.

5.5 Results and Discussion

5.5.1 CZE of rhDNase: pH Experiments

The free solution migration of rhDNase through a coated capillary in an electric field changes with hydrogen ion titration (Figure 1). These changes can be linked to the contributions of ionizable groups (Table I) [5] that are presumably exposed on the surface of the protein. At high pH (≥ 9.0) the protein possesses a high degree of negative charge and migrates quickly through the capillary toward the anode, moving as a single, relatively compact band. Between pH 9.0 and 8.0 the amino terminal becomes protonated, adding a positive charge to the protein and slowing migration toward the anode, causing the band to broaden. Between pH 8.0 and 6.0 there is little change in the ionization of protein side-chains and migration times stabilize. Below pH 6.0 the histidine residues become protonated, adding five positive charges to the protein, further slowing migration and allowing observable band broadening plus some resolution of distinct bands. At pH 4.5 the aspartic and glutamic acid residues, which make up 13% of the total amino acid composition of this protein, begin to become protonated. Below pH 5.0, the neutralization of 34 negative charges causes migration of rhDNase to slow to the extent that baseline recovery time is doubled for the resolving band profile. Only the remaining negative charges from the carboxyl terminal and the glycosylation-based sialic $(1-2 \text{ mol}^{-1})$ and phosphoric $(0.5-1.0 \text{ mol}^{-1})$ acid moieties allow for the continued slow migration of the molecule toward the anode. Furthermore, along with the decrease in mobility, additional heterogeneity is accentuated when the remaining charges on the protein are nearly neutralized in a pH-buffered solution environment that approaches the pI of rhDNase [3].

A plot of the data generated by the CZE experiments of rhDNase migration times across the pH range (Figure 2) appears similar to the theoretical hydrogen-ion titration curve of the net average charge (Z) of the protein. The theoretical net average charge is based on the amino acid composition of the protein, without consideration of structural or electrostatic interactions, and also does not include glycosylation-based charges [6]. The similarity of the theoretical and experimental pH effect curves indicates that protein migration is correlated to the net average charge of the molecule. This correlation implies that the measured electrophoretic mobilities can be used to calculate theoretical net charge, as was demonstrated for bovine serum albumin [7]. Furthermore, the similar pattern of migration and net charge indicates that most of the titratable groups in rhDNase are surface-oriented, exposed and accessible to the surrounding solution environment and applied electric field. The marked deviation from the theoretical curve observed at low pH for the measured migration time can be explained by an effect that is inherent in using coated capillaries for capillary zone electrophoresis. The increasing neutralization of acidic charges on rhDNase decreases migration until it ceases to migrate towards the detector when the protein becomes uncharged and immobile at the pI. As the pH drops below the pI the now positively charged protein migrates in the opposite direction, toward the cathode and away from the detector. Finally, protein migration through the capillary across the pH range may deviate from theoretical net charge considerations due to the occurrence of changes in hydrodynamic size of the molecule, which has been shown to undergo increases in apparent molecular weight and Stokes radii as the pH increases from 4.7–9.5 and was attributed to a progressive conformational deviation or unfolding from the spherical form [8].

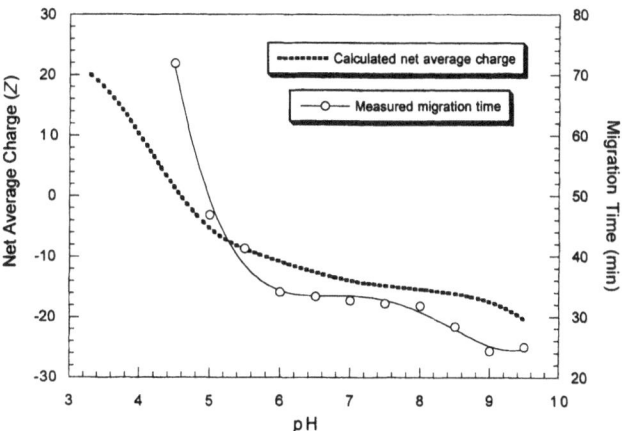

Figure 2. Theoretical and experimental pH effects on rhDNase. The theoretical hydrogen-ion titration curve (dashed line) for rhDNase was generated by the ProTitrate (version 2.0B) application program [6]. The calculation of net average charge (Z) on the protein assumes no interactions between titratable residues and does not include glycosylation-based charged residues. The experimentally determined migration times, shown in Figure 1, are plotted as a function of pH (thin line with circle markers).

5.5.2 CZE of rhDNase: Metal Ion Additions

Neutralization of net negative surface charges can also occur from association with metal counter-ions. Acidic rhDNase molecules will bind with many positively charged ions, especially under conditions of basic pH. The divalent metal cations calcium, manganese, and magnesium have been shown to have fundamental roles in the stability and activity of bovine DNase (bDNase) [4], which is nearly 80% homologous to the human form and has a similar predicted pI (5.1). The binding affinities of bDNase to calcium, manganese, and magnesium were determined to be 14, 50, and 240 micromolar, respectively, at pH 7.5, with 2–7 ions bound at both high and low affinity sites [9]. These particular divalent metal ions also offer protection against proteolytic degradation, with magnesium somewhat protective, manganese more protective, and calcium completely protective in comparison to monovalent sodium ions which offer no protection against proteolysis [10]. Measured structural changes in the circular dichroism spectra of bDNase bound to divalent metal cations also indicate that calcium was most effective, manganese somewhat effective, and magnesium had no effect on inducing changes in ellipticity [10].

The addition of divalent metal ions to a basic pH buffer has an effect similar to lowering the buffer pH on the CZE profiles of rhDNase. The addition of 0–5 millimolar (Figure 3) calcium, manganese, and magnesium at pH 8.0 causes 2–3 fold increases in the migration times of

rhDNase. However, while metal ion interactions can generally alter surface charges and thereby slow protein migration in an electric field, there also appears to be a more specific ion-affinity effect. The interaction between rhDNase and magnesium, manganese, and calcium not only slows migration but also enhances protein-associated heterogeneity in a trend that is consistent with the known binding affinities, proteolytic protection, and induced spectral changes of these divalent cations for bDNase. In contrast, additions of a monovalent metal counter-ion such as sodium have little effect on the migration times or appearance of the profiles, indicating little if any interaction with rhDNase in an electric field. Instead, the sodium chloride additions provide a control for the general effects that increasing ionic strength will have on the protein profiles. Similarly, migration time changes have also been observed to occur with protein-protein interactions, such as the bDNase-actin non-covalent complex, which maintains the bDNase profile but migrates between the widely separated free molecules [11].

The electrophoretic mobility of rhDNase (Figure 4) decreases by interaction of the protein with specific divalent metal counter-ion additions to the free solution environment. The decreasing trend in electrophoretic mobility indicates that rhDNase continues to interact with divalent counter-ions at concentrations above the one millimolar level, despite micromolar binding constants for the similar bDNase molecule, and is possibly due to the presence of the electric field in the

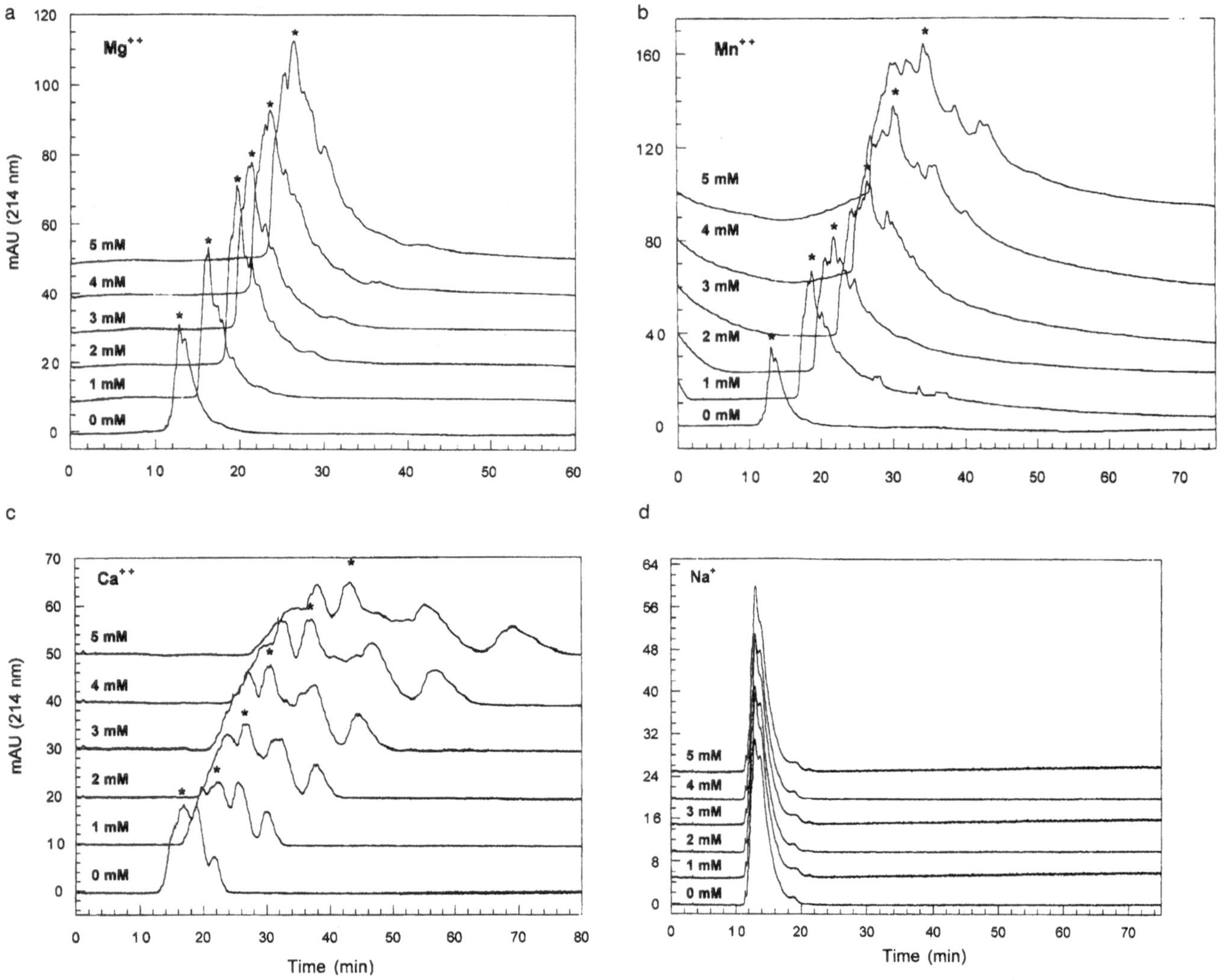

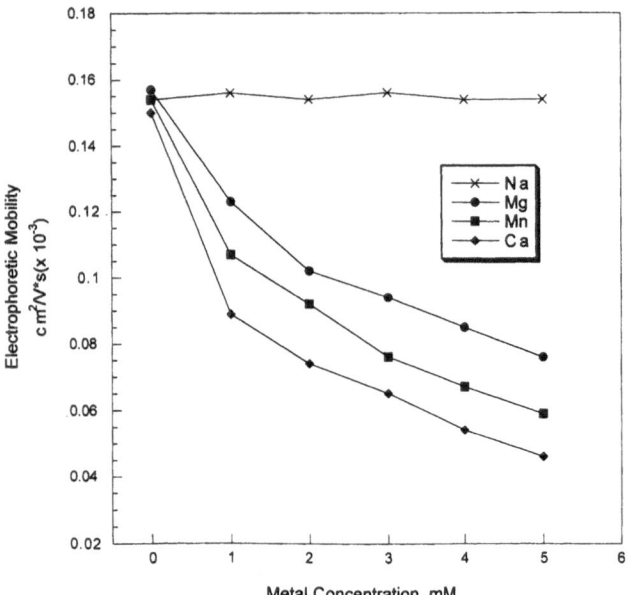

Figure 3. CZE of rhDNase with metal counter-ion additions to the buffer. Metal ions were added to a 20 mM Tris-HCl buffer at pH 8.0. The asterisks indicate the migration times used for the calculations of electrophoretic mobility shown in Figure 4.

Figure 4. Effect of metal ion additions on mobility of rhDNase. In the calculation of protein electrophoretic mobility, the effective capillary length (56 cm) is divided by the detected peak migration time (in seconds, from Figure 3), then multiplied by the inverse of the applied electric field (volts per centimeter of total capillary length = 465 V/cm) [15]. The effective electrophoretic mobility is calculated assuming no electro-osmotic flow, due to the use of coated capillaries.

CZE mode. Binding constants for bDNase-metal counter-ions were determined under equilibrium conditions by gravity flow through an open column, in which the concentration of buffer and additive ions remains the same [7]. In CZE mode, the electric field applied across the capillary drives the non-covalently associated, positively charged counter-ions toward the cathode as the protein moves toward the anode, creating an ion flow that changes the buffer composition during the separation. The continuing trend of decreasing electrophoretic mobility of rhDNase with increasing concentrations of divalent cations indicates the presence of additional negatively charged surface sites that can interact with and be neutralized by the flow of positive ions. The attainment of equilibrium conditions in the electric field will stabilize the electrophoretic mobility of the protein, when there is sufficient metal ion concentration to compensate for both the ion-flow away from the protein and the number of negatively charged sites of interaction on the surface of the protein. At acidic pH (~5), when rhDNase has fewer negatively charged sites to interact with divalent cations, electrophoretic mobility stabilizes above the 3 mM level for the addition of calcium [3]. Similarly, stability is reached in the protein-protein interaction profile of the bDNase-actin complex when the reactant concentrations (1:1 molar ratio) are at equilibrium with the complex formation [11].

The addition of counter-ions (hydrogen or divalent metal) to the free solution environment yields increasingly broad and heterogeneous profiles of rhDNase, as the protein surface carries less net charge. Neutralization of surface net charge leads to enhanced resolution of remaining charge and/or size differences. Enhanced resolution of subtle charge and/or size heterogeneity occurs when the different species have the greatest charge difference [12]. For example, with hydrogen ion titration the protein carries three negative net charges at pH 4.8 and resolves into more than two peaks, compared to the single peak at that migrates pH 8.5 when the protein has sixteen negative net charges (amino acid composition only) [6]. A possible basis for the separation is from different amounts of ions associating with and neutralizing the surface charges on rhDNase and then resolving as discreet peaks in the electropherogram. A distribution of ionic interactions between the buffer solution components and the multiple lower affinity sites of the protein would result in a heterogeneous charge-based separation; however, glycosylation-associated heterogeneity remains as an alternative source of the complex array of variant species. Although the amino acid portion of the fully deamidated glycoprotein has little or no further characterized variation, both of the asparagine-linked glycosylation sites have structures that extend outward from the surface and are heterogeneous, differing in charge and hydrodynamic size [2].

5.5.3 CZE of rhDNase: Glycan Charge and Size Heterogeneity

To account for the possible contributions of glycosylation-based heterogeneity to the resolving profiles, enzymatic digestion was used to manipulate the known associated charge and size differences. Under

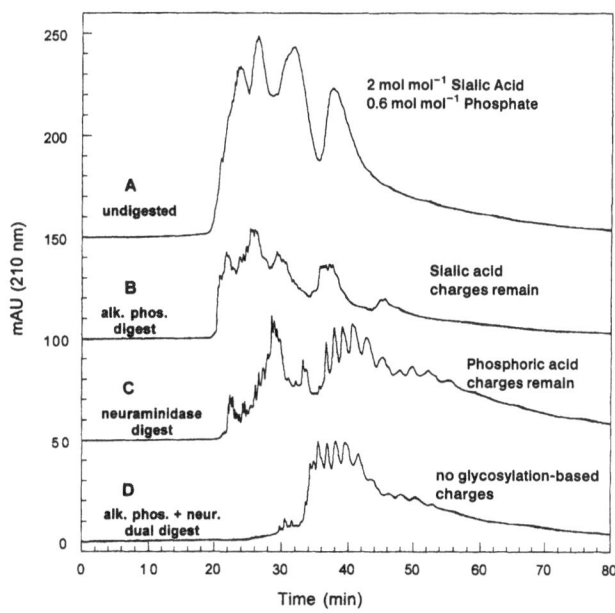

Figure 5. CZE of rhDNase digests. The separation buffer is 20 mM Tris-HCl, 3 mM CaCl₂, pH 8.0. The starting material sample (**A**, undigested) has 2 moles sialic acid and 0.6 moles phosphate per mole of protein. The sample in **B** was digested with alkaline phosphatase, in **C** with neuraminidase, and in **D** with both alkaline phosphatase and neuraminidase.

CZE conditions of a high pH buffer (20 mM Tris, pH 8.0) with added calcium (3 mM), the protein profile resolved into four broad peak areas (Figure 5A). The differences in the glycosylation of the protein in these regions were investigated by comparison of specific removal of glycosylation-based charges. Specific removal of phosphate charges by digestion with alkaline phosphatase reveals a profile similar to the starting material (Figure 5B), whereas the specific removal of sialic acid charges by digestion with neuraminidase (Figure 5C) reveals a profile significantly shifted toward the slower migrating region. This evidence indicates that the CZE profile of the original, undigested material (Figure 5A) is dominated by the sialic acid content of the glycostructures. Even though approximately 50% of the rhDNase molecules are phosphorylated [13, 14], the molar amount of sialic acid predominates by 3–4 fold (approximately 2 moles sialic acid versus 0.6 moles phosphoric acid).

Removal of all glycosylation-based charge differences by digesting the protein with both neuraminidase and alkaline phosphatase (Figure 5D) collapses the profile into a broad zone of lower mobility, with an additional level of heterogeneity profiled across the top of this region. Although these results indicate that the original broad profile of associated rhDNase variants are indeed related to glycosylation-based charge differences,

the observed remaining microheterogeneity after digestion with both neuraminidase and alkaline phosphatase could arise from a charged-based distribution of ionic interactions with the solution components. However, there also remains glycosylation-based size heterogeneity that may account for the additional resolution. Following enzymatic removal of the charged moieties, rhDNase glycoforms can range in mass (and hydrodynamic size) from approximately 31.7 kDa (for pentamannose structures at both sites) to approximately 34.2 kDa (for fucosylated tetraantennary structures at both sites).

Complete deglycosylation of the protein by digestion with PNGase F (Figure 6) produced a profile of rhDNase as a single sharp band, with no apparent charge or size variation associated with the remaining protein. The single band indicates that the heterogeneous profile of rhDNase is the result of the heterogeneous glycosylation structures that vary in both charge and size, and not from a distribution of neutralized surface-charged sites. Deglycosylated rhDNase, without the carbohydrate-associated sialic and phosphoric acid moieties, becomes a less acidic molecule and therefore should move more slowly through the capillary compared to the native, glycosylated protein. The reduction in hydrodynamic size allows the deglycosylated protein band to move through the applied electric field in the capillary more quickly than the intact

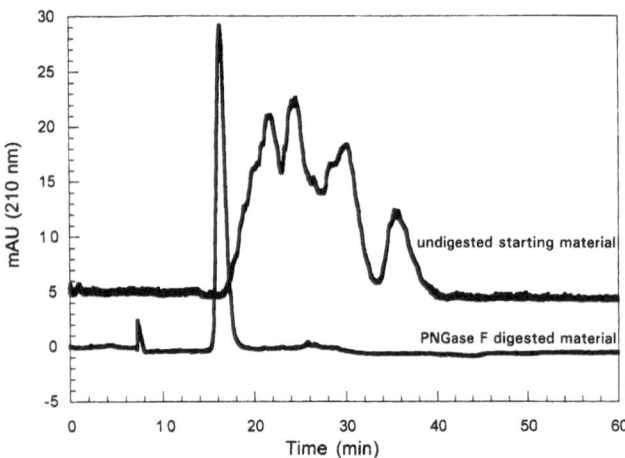

Figure 6. CZE of native and deglycosylated rhDNase. The separation buffer is 20 mM Tris-HCl, 3 mM CaCl$_2$, pH 8.0. The deglycosylated rhDNase was digested with PNGase F.

glycoprotein. Additionally, the enzymatic reaction that releases the asparagine-linked oligosaccharide structures converts these asparagine residues into aspartic acid, adding two negative charges. Finally, the loss of glycosylation structures could expose previously shielded charged surface sites to the surrounding electrical field, altering the electrophoretic mobility of the deglycosylated form.

5.6 Conclusion

These experiments describe the analysis of rhDNase, a complex glycoprotein, by capillary zone electrophoresis. This work demonstrates that the migration of rhDNase through a capillary coated with a hydrophilic polymer is strongly influenced by the composition of the free solution environment. Changes in either the hydrogen ion or divalent metal ion concentration of the buffer will alter the net surface charge of the rhDNase molecule, causing changes in mobility in the presence of an electric field. Specifically decreasing net protein surface charges, which occurs when the buffer pH nears the isoelectric point or with the addition of specific divalent metal cations to the buffer, also leads to enhancement of variant resolution, since neutralization of the bulk of rhDNase net charges accentuates remaining charge and/or size heterogeneity. The CZE separation profile of rhDNase reveals a complex array of closely associated variant structures that are shown to be based primarily on the charged sialic acid content of the glycosylation structures, with a remaining level of heterogeneity based on glycan size differences.

5.7 Acknowledgements

The authors would like to thank Namita Nayak for the sialic acid analysis, and Stacey Ma and Fred Jacobson for review of the work.

5.8 References

[1] Shak, S.; Capon, D.J.; Hellmiss, R.; Marsters, S.A.; Baker, C.L. *Proc. Natl. Acad. Sci.* U.S.A. **1990**, *87*, 9188–9192.
[2] Quan, C.P.; Cacia, J.; Frenz, J.; O'Connor, J.V. *Pharm. Sciences* **1997**, *3*, 53–57.
[3] Felten, C.; Quan, C.P.; Chen, A.B.; Canova-Davis, E.; McNerney, T.; Goetzinger, W.K.; Karger, B.L. *J. Chromatogr. A* **1999**, *853*, 295–308.
[4] Price, P.A. *J. Biol. Chem.* **1975**, *250*, 1981–1986.
[5] Cantor, C.R.; Schimmel, P.R. "Biophysical Chemistry, part I, The Conformation of Biological Macromolecules", W.H. Freeman and Co., San Francisco, **1980**, pages 41–53.
[6] Shire, S.J. *Biochemistry* **1983**, *22*, 2664–2671.
[7] Menon, M.K.; Zydney, A.L. *Anal. Chem.* **1998**, *70*, 1581–1584.
[8] Lizárraga, B.; Sánchez-Romero, D.; Gil, A.; Melgar, E. *J. Biol. Chem.* **1978**, *253*, 3191–3195.
[9] Price, P.A. *J. Biol. Chem.* **1972**, *247*, 2895–2899.
[10] Poulos, T.L.; Price, P.A. *J. Biol. Chem.* **1972**, *247*, 2900–2904.
[11] Carter, L.K.; Christopherson, R.I.; dos Remedios, C.G. *Electrophoresis* **1997**, *18*, 1054–1058.
[12] Martin, L.M. *J. Chromatogr. B* **1996**, *675*, 17–25.
[13] Frenz, J.; Quan, C.P.; Cacia, J.; Democko, C.; Bridenbaugh, R.; McNerney, T. *Anal. Chem.* **1994**, *66*, 335–340.
[14] Cacia, J.; Quan, C.P.; Pai, R.; Frenz, J. *Biochemistry* **1998**, *37*, 15154–15161.
[15] Heiger, D.N. "High Performance Capillary Electrophoresis – An Introduction", Hewlett-Packard Company, France, **1992**, page 23.

Recent Advances in Capillary Isoelectric Focusing

T. Wehr[1] / R. Rodriguez-Díaz[2] / M. Zhu[3]

[1] LC Resources, Inc., 2930 Camino Diablo, Suite 110, Walnut Creek, CA 94596, USA
[2] Dynavax Technologies Corp., 1717 Potter St. Suite 100, Berkeley, CA 94710, USA
[3] Bio-Rad Laboratories, 2000 Alfred Nobel Drive, Hercules, CA 94547, USA

6.1 Introduction

Capillary isoelectric focusing (cIEF) combines the high resolving power of conventional gel isolectric focusing with advantages of capillary electrophoresis (CE) instrumentation. As in gel IEF, proteins are separated according to their isoelectric points (pIs) in a stable pH gradient formed by ampholytes under the influence of an electric potential. When performed in the capillary format, the use of small-diameter capillaries with efficient dissipation of Joule heat permits the application of high field strengths for rapid focusing. The reduced convection in capillaries removes the requirement for gels. The use of UV-transparent fused silica capillaries enables direct on-tube optical detection of focused zones without the requirement for staining. The focusing effect provides resolving power up to 0.02 pI. A unique feature of cIEF with on-tube detection is the requirement for mobilization of the focused protein zones across the detector window.

The extraordinarily high resolution and the automation features of cIEF would seem to make it an ideal tool for characterizing biopharmaceutical proteins. However, the technique requires the control of a large number of variables. Poor understanding of the cIEF process limited its application as a robust analytical technique in industrial settings. However, improvements in methodology and capillary quality have altered this picture in recent years, and cIEF is currently being employed in development and quality control environments for characterizing protein microheterogeneity and stability.

Several reviews of cIEF have been published [1–3]. This contribution will focus on more recent developments, and particularly on advances in the last four years.

6.2 Theory and Practice of cIEF

6.2.1 Ampholyte Selection

Ampholyte composition determines the range of the pH gradient and the resolution within that range. A broad range ampholyte mixture [e. g. 3–10] will be useful for separation of complex mixtures with widely differing pI values, or for screening a protein with unknown pI. For high resolution within a narrow pH range, narrow-range ampholytes which generate gradients over as little as one pH unit can be used. However, these should include small amounts of broad-range ampholytes to span the gap between the narrow-range ampholytes and the anolyte and catholyte solutions. Resolution within a given range is a function of the number of ampholyte species in the mixture: the greater the number of species, the higher the resolution. If ampholyte preparations from a single supplier do not provide sufficient resolution, ampholytes from a variety of suppliers can be blended to increase the number of ampholyte species [4]. One unfortunate characteristic of ampholytes is their strong absorbance in the low UV.

All commercial ampholytes were developed for conventional gel IEF in which this property is irrelevant. In cIEF, a detection wavelength must be selected at which there is minimal interference from the ampholytes but satisfactory signal for proteins. A wavelength of 280 nm is typically used; although protein absorbance at 280 is typically 50-100-fold less than in the low UV, the very high protein concentration in focused zones compensates for the loss in signal.

Since the detection point is located at some distance from the capillary end, proteins which focus between the detection window and the capillary end will not be detected in two-step cIEF. This problem can be resolved by including a spacer such as N,N,N',N'-tetramethylethylenediamine (TEMED) in the protein + ampholyte mixture [5]. At the completion of focusing, the spacer occupies this "blind" segment of the capillary, and all proteins are detected during mobilization.

The use of spacers to resolve proteins with very similar isoelectric points has been proposed by Righetti et al. [6]. For example, β-alanine (pI 6.90) was added at a concentration of 100 mM to a pH 6–8 ampholyte mixture to resolve hemoglobins A and F. Similarly, β-alanine (300 mM) + 6-aminocaproic acid (330 mM, pI 8.04) were added to the same ampholyte blend to resolve hemoglobin A from hemoglobin A_{1C}. The spacers served to locally flatten the pH gradient in the region of the analytes and achieve their resolution.

0009-5893/00/02   45-14 $ 03.00/0

6.2.2 Mobilization Techniques

In contrast to conventional gel isoelectric focusing in which focused proteins are fixed and visualized with a dye, capillary IEF requires a method of mobilize the protein zones past a fixed detection point. This can be performed while focusing is in progress, in which case the technique is termed single-step cIEF. Alternatively, mobilization can be performed as a separate step after focusing; this is termed two-step cIEF. In single-step cIEF, the mobilization force can be hydraulic (pressure, vacuum, or gravity) or electroosmotic. In two-step cIEF, the mobilization force can be hydraulic or electrophoretic.

Single-Step cIEF

In the original approach for single-step cIEF described by Mazzeo and Krull [7], the sample was mixed with ampholytes and injected into an uncoated silica capillary. Focusing occurred during transport of the mixture to the cathode by EOF. A refinement of this approach employed capillaries coated to maintain a constant level of EOF [8]. Tang and Lee [9] have shown that suppression of electroosmotic flow in uncoated capillaries filled with high concentrations of ampholytes prevents EOF mobilization. An alternative approach to single-step cIEF was described by Thormann et al. [10]. In this technique, the sample + ampholyte mixture was injected as a plug into a capillary prefilled with catholyte or catholyte mixed with alkylated cellulose. Gradient formation and protein separation occurred as the sample + ampholyte zone was transported towards the detection point by EOF. Disadvantages of EOF-driven mobilization include the requirement for extensive capillary conditioning steps between analyses, the need for careful adjustment of EOF with polymer additives, and broadening of acidic protein peaks.

Single-step cIEF Using Anionic Coated Capillaries

Some of the limitations of EOF-driven mobilization were eliminated by the use of anionic coated capillaries. Whynot et al. [11] used a modification of the Hjertén procedure [12] to copolymerize acrylamide and sodium 2-acrylamido-2-methypropanesulfonate. The net anionic charge of the resultant covalent polymeric coating could be adjusted by varying the ratio of neutral and acidic monomers. The

strong acid functions of the coating generated electroosmotic flow which was pH-independent in the pH range of 3 to 9. Separations performed with this coating were faster and required only a simple water rinse between injections. However, significant broadening of acidic protein peaks was still observed.

Single-step cIEF Using Segmented Injection

A variation on the single-step protocol for cIEF has been described by Kilár et al. [13]. In this report, sample and ampholytes were injected as separate zones, with the sample zone bracketed by ampholyte zones. The two ampholyte zones may be of differing composition or volume. Analysis parameters were optimized using synthetic aminomethylated nitrophenols, and the method was applied to separation of hemoglobin variants. Performance of this method compared to nonsegmented single-step cIEF was similar in terms of resolution and reproducibility. It offers the advantage of minimizing the interaction of proteins and ampholytes prior to injection, and of detecting components which lie outside the pH range of the ampholyte gradient.

Two-Step cIEF with Electrophoretic Mobilization

Electrophoretic mobilization is typically accomplished by replacing the catholyte with a mobilization solution following the initial focusing step. The original technique described by Hjertén and Zhu [14] employed an alkaline sodium chloride solution or phosphoric acid. A later refinement used a zwitterion solution as the mobilizer [5], which provided better mobilization of acidic proteins such as phycocyanins (pI 4.45–4.75) at the distal end of the pH gradient. A more recent study evaluated different salts as mobilizers and reported that sodium tetraborate provided improved resolution compared to sodium chloride, sodium acetate, or dibasic sodium phosphate [15]. These authors also demonstrated a simple technique for performing gravity mobilization.

Manabe et al. [16] compared sodium chloride with strong and weak acid solutions as mobilizers in two-step cIEF. Phosphoric acid produced a flattening of the gradient in the pH 6–7 region which was attributed to conversion of HPO_4^{2-} ions to $H_2PO_4^-$ during mobilization. Acetic acid provided best resolution of human plasma proteins, particularly for

acidic species. However, the use of low-mobility organic anions almost doubled the mobilization time.

Two-Step cIEF with Hydrodynamic Mobilization

Hydrodynamic mobilization of focused zones has been accomplished by application of pressure on the capillary inlet, by application of vacuum at the capillary outlet, or by siphoning (gravity mobilization). Gravity mobilization has been accomplished by elevating the inlet vial or by reducing the volume of liquid in the outlet vial. In all forms of hydrodynamic mobilization, zone broadening arises from the parabolic flow profile introduced by the hydrodynamic force applied.

Minárik et al. [17] presented a theoretical treatment of peak dispersion in pressure-driven mobilization which accounted for contributions from flow profile and adsorption effects. Theoretical predictions were tested by experimental measurements of plate height (H) using pressure mobilization of focused myoglobin in uncoated capillaries. As expected, plate height increased with increasing mobilization pressure. The contribution of protein adsorption to dispersion was assessed by varying the amount of a dynamic coating material, hydroxypropylmethylcellulose. As also predicted by theory, the slopes of H vs. mobilization velocity plots were reduced by increasing the amount the dynamic coating. The authors also proposed that since adsorption of charged protein to the silica surface should be pH dependent, there should exist a gradient of adsorption along the pH gradient. The influence of mobilization direction on dispersion supported this prediction: the slope of the H vs. velocity plot for displacement from the cathodic side was significantly less than that for anodic displacement.

6.2.3 Internal Standards for cIEF

The large number of variables which need to be controlled in cIEF has limited the reproducibility of the technique and prevented it from being accepted as a reliable analytical tool until recently. One approach which has greatly improved the quality of the data in cIEF separations is the use of internal standards. Although protein standards may be run separately to generate calibration plots of pI vs. mobilization time, they are usually unsatisfactory for use as internal standards. Proteins may not exhibit good stability or batch-to-batch purity. Many proteins contain degradation products, variants, and isoforms which give rise to multiple peaks which interfere with data interpretation. Recently two types of compounds have been described for use as internal standards in cIEF, and one of them is commercially available.

A family of colored ampholytes have been synthesized by Slais and Friedl [18] for use as internal standards in cIEF. These are substituted aminomethylphenol compounds which are highly water-soluble, stable, and exhibit strong absorbance at UV and visible wavelengths. Over two dozen of these pI markers have been synthesized with pI values ranging from 5.3 to 10.4. Their strong UV absorbances permit them to be added to samples at low concentration, and the wide range of available pI values usually allows selection of marker pairs which bracket the pI values of analyte proteins. Virtually no interaction was observed between a mixture of the pI markers and the protein alcohol dehydrogenase. A subset of these markers is available commercially from Bio-Rad Laboratories (Hercules, CA) under the trade name BioMark.

Kobayashi et al. [19] prepared dansyl derivatives of both peptides and ampholytes for use as internal markers to evaluate pH gradient formation of a number of commercial ampholyte preparations. Pressure-driven mobilization in a neutral hydrophilic coated capillary was used in this study. The dansyl derivative of the tripeptide GGH was purified by reversed phase HPLC and a pI of 5.24 was estimated by comparison to mobilization times of protein standards. Protamine was digested with chymotrypsin, reacted with dansyl chloride, and two derivatives were purified by preparative isoelectric focusing. The two derivatives migrated close to ribonuclease and β-lactoglobulin A, respectively. The pH range and resolution of five commercial wide-range amphoyte mixtures were compared using the dansyl peptides.

6.2.4 Desalting Procedures for cIEF

The presence of salt in the sample can have disastrous consequences for the cIEF process. During the focusing step, the presence of high mobility salt ions generate high initial currents which cause Joule heating and loss of resolution. As salt ions exit the capillary, they are replaced by segments of acid or base at the anodic and cathodic ends of the capillary. These segments compress the pH gradient, compromising resolution and increasing the risk of protein precipitation. Since most biological samples contain salt or buffer ions, a desalting procedure is usually necessary for most samples. If the protein is present in high concentration and the sample ionic strength is relatively low, simple dilution with the ampholyte mixture may suffice. For samples with high salt content or low protein concentration, off-line desalting procedures such as dialysis or ultrafiltration can be used. However, these steps add cost in terms of time and materials to the analysis. Three on-line desalting techniques have been reported.

An on-line desalting method using ampholyte replacement was described by Liao and Zhang [20]. In this technique a solution of ampholytes was titrated to pH 4.0 with HCl and used as anolyte in the desalting step, and a solution of amphoytes titrated to pH 11.0 with NaOH was used as catholyte. After injection of the sample + ampholyte mixture, an electric field was applied at constant current (to minimize Joule heat during desalting). Under these conditions, the titrated ampholytes were charged and migrated into the capillary to replace the exiting salt ions. Following desalting, the titrated ampholyte solutions were replaced with conventional anolyte and catholyte solutions, and cIEF focusing + mobilization steps were performed as usual. Although this technique is useful when sample limitations prevent off-line desalting, it is limited by variations in the ampholyte distribution in the final pH gradient as a function of variation in sample salt content.

Clarke et al. [21] employed voltage ramping as an on-line technique to remove salt prior to focusing. After the capillary was filled with the sample + ampholyte mixture, the voltage was increased linearly from 0–10 kV over 6 min. The initial low voltage allowed high-mobility salt ions to exit the capillary without excessive Joule heating. The desalting procedure was optimized with mixtures of standard proteins containing up to 100 mM final salt concentration. Slight increases in mobilization times were observed which was attributed to gradient compression. Capillaries coated internally with poly(vi-nylalcohol) were used in this study, and pressure mobilization was compared to electrophoretic mobilization. The desalting technique was applied to cIEF analysis of human blood and cerebrospinal fluid samples. A disadvantage of on-line electrophoretic desalting is the need to optimize the voltage ramp time and range for a given sample salt concentration. It should also be noted that this procedure serves to remove high-mobility salts without excessive heating, but does not solve the problem of gradient compression and exposure of the capillary to extremes of pH.

6.2.5 Minimizing Protein Precipitation

Protein precipitation is a common problem in all forms of isoelectric focusing. The focusing process brings protein molecules together at high concentrations under isoelectric conditions. The focusing process also strips any salt associated with the protein. These three characteristics of IEF greatly increase the tendency for proteins to aggregate and fall out of solution. In gel IEF, precipitation is evidenced by smearing of protein bands and loss of resolution. In cIEF, precipitation is evidenced by fluctuating current, appearance of spikes on the baseline, and variable mobilization times. In worst cases, precipitation blocks the capillary and current falls to zero. The risk of precipitation can be minimized by adding nonionic surfactants (Triton X-100, Brij, Tween) organic modifiers (glycerol, ethylene or propylene glycol) or chaotropes (urea) to the sample [5, 22]. Conti et al. [22] improved protein solubility in cIEF with the addition of high concentrations of polyols such as 20–40% sucrose, sorbose or sorbitol in combination with high concentrations of zwitterions (e. g. 200 mM taurine, 500 mM nondetergent sulfobetaines, 1 M bicine, or 1 M CAPS). Zwitterions act by increasing the surface tension of water, forcing the protein into a "superhydrated" state. A particularly intractable problem with precipitation of a glycopeptide antibiotic was successfully solved with a mixture of 6 M urea + 10% trifluoroalcohol.

6.3 cIEF in Microchannels

The development of microscale analysis systems based on chip technology is a ra-

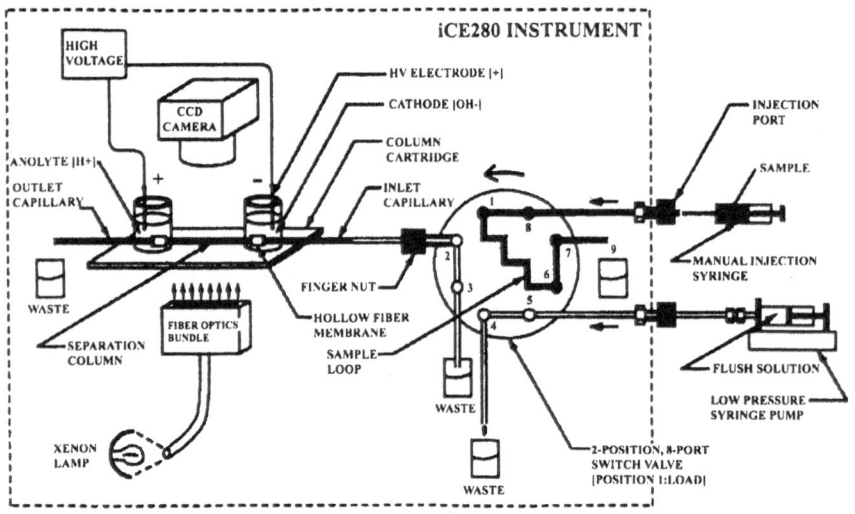

Figure 1. Block diagram of imaging capillary isoelectric focusing system. Reprinted with permission from International Scientific Communications, Inc. and Convergent BioScience Ltd.

pidly expanding field. Chip-based microchannel devices are envisioned as tools for high-throughput clinical diagnostics and for rapid screening of drug candidates in the pharmaceutical industry. Separations in microchip devices are typically electrically-driven, and IEF may be the mode of choice where high-resolution separation of proteins is required.

Hofmann et al. [23] investigated conditions for isoelectric focusing in microchannels as part of a development effort to use cIEF as a detection method for fluorescent peptides employed as probes in a multiplex diagnostic assay system. Although preliminary studies with proteins using conventional CE instrument indicated that 2-step cIEF with electrophoretic and hydrodynamic mobilization provided best resolution and reproducibility, single-step cIEF with EOF mobilization was selected due to compatibility with the chip format. The microchannel device consisted of a 200 μm wide, 10 μm deep channel of 7 cm length etched in planar glass. Reservoirs consisted of pipet tips inserted into holes in the cover plate at each end of the channel. Laser-induced fluorescence was used to detect Cy-5 labeled peptides. The channel was filled with a mixture of ampholytes + peptides containing 40% glycerol, then focusing and mobilization performed concurrently in the presence of EOF. The high surface-to-volume ratio provided good dissipation of Joule heat, allowing field strengths of 1 kV/cm. Best sensitivity and resolution was obtained by positioning the detection point midway along the separation distance.

In a report describing progress in the construction of microscale devices using photoablated polymers, Rossier et al. [24] proposed the use of IEF for protein separations in microchannels. The proposed design consisted of a network of microchannels connecting reservoirs containing immobilized ampholytes at different pH values. The design would permit entrapment of proteins of interest in reservoirs corresponding to their isoelectric points. The concept was illustrated by polymerization of a polyacrylamide gel within a microchannel; the pH of the gel was set to 5.4 by the inclusion of a mixture of immobilized ampholytes. A mixture of cytochrome C and β-lactoglobulin was placed in the anode reservoir and voltage applied. The immobilized ampholytes formed an isoelectric sieve, allowing only the cytochrome C to migrate to the cathode reservoir.

6.4 Imaging cIEF

Use of a conventional CE instrument with single-point on-tube detection requires a mobilization step to transport focused zones past the monitor point. This requirement has several drawbacks for isoelectric focusing. Mobilization lengthens the analysis time, reducing sample throughput. In two-step cIEF, the high voltage must be turned off for a sufficient time to replace the catholyte or anolyte with the mobilizing solution; this interruption permits some analyte diffusion which could compromise resolution. In electrophoretic and EOF mobilization, variations in mobilization velocity can lead to uneven resolution along the pH gradient. In hydraulic mobilization, laminar flow effects can compromise resolution.

A novel approach to cIEF which eliminates the problems of mobilization has been developed by Pawliszyn's group [25, 26]. This technique, termed imaging cIEF, employs a CCD imaging detector to monitor the entire length of the capillary during the focusing process, thus obviating the need for mobilization. A commercial instrument for imaging cIEF has recently been introduced [27, 28].

A schematic of the imaging cIEF system is presented in Figure 1. A short 5 cm fused silica capillary, stripped of its outer polyimide cladding and coated internally with polyacrylamide or a fluorocarbon to eliminate EOF, is housed in a cartridge. The two ends of the capillary are attached to lengths of hollow dialysis fiber membrane which are inserted into the electrolyte reservoirs. The fibers serve to isolate the capillary contents from the reservoirs but allow free passage of anolyte and catholyte ions. Light from a Xenon lamp is distributed by an optical fiber bundle along the capillary length and transmitted light is imaged onto the CCD detector. The capillary inlet is interfaced with an 8-port valve connected to an autosampler and an infusion pump. The sample can be introduced into the sample loop manually or by an HPLC autosampler. The valve is then rotated to the inject position and the sample is displaced into the capillary by the infusion pump. Voltage is applied and the focusing process can be monitored in real time; at the completion of focusing, the zone profile can be recorded.

As in gel IEF and conventional cIEF, protein precipitation can be a problem in imaging cIEF. In such cases, additives such as glycerol and urea can be incorporated in the sample mix. In addition, the focusing process can be monitored to determine the onset of precipitation, and the analysis time adjusted to avoid precipitation. In cIEF, all segments of the pH gradient are exposed to the same focusing time; this is in contrast to conventional cIEF in which late-mobilizing proteins are exposed to focusing conditions for a longer period and are therefore at greater risk for precipitation,

The cIEF imaging technique has been applied to hemoglobin variants, glycoproteins, monoclonal antibodies, and peptides [28], with excellent resolution and reproducibility.

6.5 cIEF-Mass Spectrometry

Two-dimensional separation systems in which the separation selectivity of the two dimensions is orthogonal offer the highest resolving power for complex mixtures. The classic example is 2D gel electrophoresis in which the first dimension (IEF) is based on pI and the second dimension (SDS-PAGE) is based on molecular size [29]. To a first approximation, the total resolution of the technique is the product of the band capacity of each dimension, and 2D gel electrophoresis has been used to resolve as many as several thousand proteins from biological fluids or extracts. In the developing field of proteomics, 2D gel electrophoresis is the standard technique for isolating proteins for structural analysis and for comparing protein expression patterns. The utility of proteomic studies in identification of drug targets and candidate protein therapeutics is creating a need for fast, high-throughput protein mapping. Unfortunately, 2D gel electrophoresis is slow, labor intensive, and semi-quantitative at best. To date, efforts to automate 2D gel electrophoresis have been unsuccessful. Coupling cIEF to mass spectrometry promises to provide 2D separations of proteins in an automated format with short analysis times. cIEF-MS has been reviewed recently [30].

On-line coupling of cIEF and electrospray mass spectrometry was first described by Tang et al. [31]. In this study, two-step cIEF with electrophoretic mobilization was employed. Focusing was performed with the catholyte reservoir placed in the electrospray interface. Following the focusing step, the catholyte reservoir was removed and mobilization of the focused zones into the electrospray source was initiated by infusion of a sheath liquid composed of methanol:water:acetic acid. Initial experiments using direct infusion of protein-ampholyte mixtures were performed to investigate the effects of ampholytes on protein ionization. It was observed that the presence of ampholytes suppressed protein ion intensities and decreased the net charge of the protein ions. This was attributed to charge neutralization caused by ampholyte-protein interaction, causing a shift in the charge distribution of the protein mass spectra to higher m/z values. Results with on-line cIEF-MS analyses indicated that the high zone concentrations provided by the focusing process improved detection limits by two orders of magnitude compared to CZE-MS.

A later modification of the technique used combined electrophoretic and gravity mobilization to compensate for moving boundary effects caused by migration of sheath liquid acetate ions into the capillary [32]. The modified cIEF-MS procedure was used to resolve glycoforms of bovine serum transferrin. The first cIEF dimension resolved di-, tri-, and tetrasialo-transferrins, while the second ESI-MS dimension resolved additional variants in each sialotransferrin which differed in molecular weight. In a similar study [33], mono- and disphosphoalbumin were resolved in the cIEF dimension, and glycoforms within the charge variants were resolved by ESI-MS. The same group used cIEF-ESIMS to identify a growth hormone-glutathione-S-transferase fusion protein in cell extracts of bacteria expressing the recombinant protein [34]. In this work, cIEF-MS could resolve approximately 100 proteins. Recently, this cIEF-MS approach was applied to analysis of standard proteins using time-of-flight MS [35].

The system described by Tang et al. [32] was limited by the need to reposition the capillary between focusing and mobilization, by ampholyte suppression of protein ion intensity, and by competition from sheath liquid ions in the electrospray process. Some of these limitations were circumvented by Lamoree et al. [36] by incorporating an on-line microdialysis device (MD) between the separation capillary and a transfer capillary connected to the ESI source. The MD consisted of a short section of hollow fiber dialysis tubing connecting the ends of the separation and transfer capillaries. The hollow fiber was sealed into a chamber through which 2% acetic acid was infused. Acetic acid (2%) was used both as catholyte and as the ESI sheath liquid. After focusing, protein zones were mobilized into the mass spectrometer by pressure. The MD removed sufficient amounts of ampholyte to prevent fouling of the ESI source and interference with the electrospray process. The MD was later modified to couple the separation and transfer capillaries by an axial dialysis chamber separated from the dialysis liquid by a flat dialysis membrane [37]. This modified cIEF-MD system was coupled to and ESI-magnetic sector MS. Yang et al. [38] also performed cIEF-ESIMS using a hollow-fiber microdialysis junction and pressure mobilization. The microdialysis liquid (10% acetic acid) served as the anolyte and provided a pro-

ton source for ionization, obviating the need for a sheath liquid. In all three studies, anodic mobilization was used so that no capillary repositioning was required between the focusing and mobilization steps. In the resultant electropherogram, acidic proteins appeared before basic proteins.

Fourier transform-ion cyclotron resonance mass spectrometry (FTICR-MS) can provide extraordinarily high mass resolution and accuracy, allowing unambiguous determination of protein charge and mass from a single charge state. Severs et al. [39] interfaced cIEF with ESI-FTICR-MS using two-step cIEF in which mobilization was accomplished by cathodic pressurization using a methanol-water-acetic acid sheath liquid. The technique was used to obtain mass spectra of carbonic anhydrase in a red blood cell lysate in the presence of a 100-fold excess of hemoglobin. Recent refinements of the technique enabled identification of proteins in bacterial cell lysates [40]. To improve sensitivity and mass accuracy, cells were grown in media depleted of rare isotopes. Cell lysates were subjected to a dual microdialysis technique to select proteins in the 3–60 kDa range. Following on-line separation of lysates by cIEF with gravity mobilization, protein masses were determined from isotopic distributions of single charge states for proteins < 30 kDa or deconvolution of charge-state distributions for larger proteins. Presentation of the data in a simulated 2-D gel format exhibited ~900 protein "spots," each with a unique mass. Good correlation was observed between the location of phosphoglycerate kinase mass on the 2-D cIEF-FTICR-MS map and the location of the protein spot on an actual 2-D gel. Preliminary experiments with protein standards demonstrated the use of cIEF-FTICR-MS[n] to obtain sequence information for unambiguous protein identification.

In a novel application of cIEF-MS, capillary isoelectric focusing was used by Michalke and Schramel [41] to resolve organic selenium complexes, which were identified by monitoring the selenium signal using inductively-coupled plasma mass spectrometry (ICP-MS). After focusing in a pH 2–10 gradient, capillary contents were pressure-mobilized into the ICP-MS nebulizer. The technique was used to identify selenocystamine, selenocystine and selenoglutathione in human milk, and selenocystine and selenoglutathione in human serum. cIEF-ICP-MS

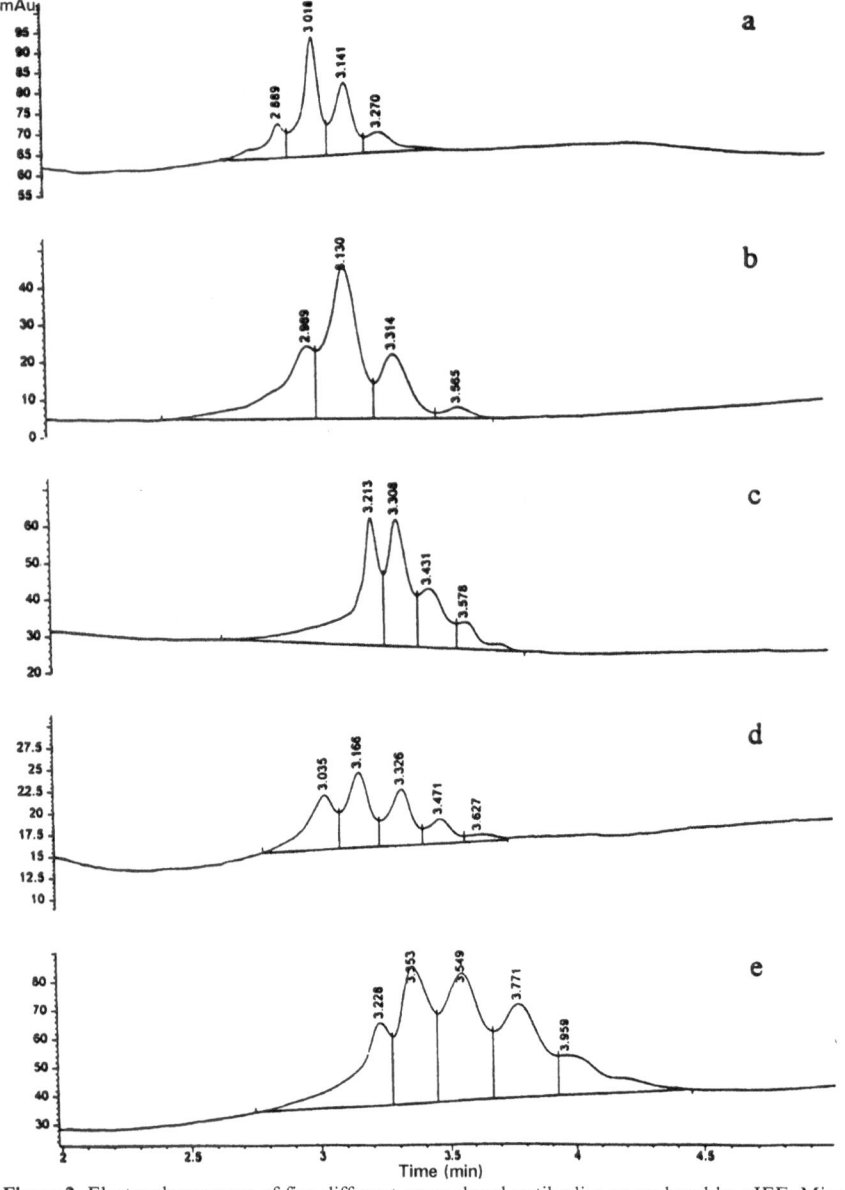

Figure 2. Electropherograms of five different monoclonal antibodies as analyzed by cIEF. Migration time is indicated on top of each peak. Samples were analyzed using various ampholyte compositions and voltages. Although the differences in analysis conditions compromises any comparison in migration time or peak areas, the technique clearly shows that each of the samples is composed of various isoforms. Reproduced from reference 44 with permission.

bohydrate content can strongly affect the biological activity, clearance, specificity, solubility and stability of an antibody. These effects are the result of changes in protein hydrophobicity, charge, mass, and/or conformation. The consistency of MAbs arises from their production using hybridoma cell lines or rDNA techniques, and cIEF can be used to monitor MAb stability and mutations.

Capillary IEF has also been used in the analysis of fragments obtained from proteolysis of MAbs. Murine MAbs (human anti-mouse antibodies or HAMAs) can cause the formation of an unwanted immune response in patients when applied in vivo [42]. One approach to minimize this effect is by partially "humanizing" the antibody by replacing the murine Fc with a human Fc. Another method is to use only the Fab or F(ab')2 fragment of the molecule. Both approaches take advantage of the fact that the location of the immunogenic properties of the IgG are predominantly located on the constant fraction (Fc) of the antibody. Antibody proteolysis may also generate fragments that show improved stability, and better general pharmacokinetics as compared to the parent IgG.

Earlier literature reports employed cIEF as a qualitative technique for protein characterization. However, cIEF techniques have now developed to the point that a cIEF method has been fully validated [43]. Method validation for quality control release of biotechnology products requires intensive testing of the performance of the technique. Only those methods that produce reliable, reproducible data become validated. In this section we will review examples of validation of cIEF methods for characterization of monoclonal antibodies.

Lee [44] developed a rapid single-step cIEF method using an uncoated fused-silica capillary to analyze various preparations of MAbs. In this feasibility study, five different monoclonal antibodies were analyzed (Figure 2) using slightly different ampholyte preparations. In all instances, multiple peaks were obtained, revealing a number of different species ranging form four to as many as six. For pI determination, antibody samples were spiked with a small volume of protein standard (horse myoglobin, pI 7.2, bovine carbonic anhydrase, pI 5.4, and 5.9, and/or human carbonic anhydrase, pI 6.6). In this method, a 32 cm (effective length 9 cm) × 50 μm I. D uncoated capillary was used. Hydroxyl-

provided better resolution and enhanced sensitivity in comparison to CZE-ICP-MS.

6.6 Capillary Isoelectric Focusing of Monoclonal Antibodies

Monoclonal antibodies (MAbs) have found widespread use as a diagnostic and therapeutic tool in an increasing number of clinical indications. When analyzed by IEF, these monoclonal antibodies often exhibit microheterogeneity, frequently yielding a significant number of discrete electrophoretic peaks. This microhetero-

geneity is due predominantly to post-translational events, particularly glycosylation. In-process purification, as well as after-storage deamidation of glutamine and/or asparagine residues can also increase the number of molecular species observed. The high resolving power of cIEF can be used to monitor protein stability, as well as freeze/thaw cycle changes. For these reasons, cIEF is well suited for characterization and it is also a stability indicating technique.

The demands for consistency in production of therapeutic antibodies, and those used in the immunodiagnostics industry are constantly pushing the limits of analytical techniques. Differences in car-

propylmethyl cellulose was added to lower the electroosmotic flow to allow sufficient time for the proteins to focus before reaching the detector. Separations were initially attempted in the longer portion of the capillary with or without TEMED as a basic extender. But as often occurs with cIEF in uncoated capillaries, this configuration gave excessive peak broadening due to the extensive interaction of the sample and the capillary wall. Reverse polarity without TEMED using the short end of the column reduced this problem. To increase the reproducibility of the assay, a daily conditioning protocol consisting in cycles of water, 1 M NaOH, water, 1 M HCl, and 0.4% HPMC was performed. In-between runs the column was conditioned with 1 M or 0.1 M NaOH, then 0.4% HPMC prior to sample injection. The ampholytes used for non-denaturing cIEF consisted of Pharmalyte (pH range 5–8) 6% v/v in 0.5% HPMC, or 0.8% HPMC. The author did not elaborate on the purpose of the ampholyte formulation change. For denaturing cIEF a solution containing Pharmalyte (pH range 5–8) at 10% v/v in 7 M urea and 0.3% HPMC was used. The sample was prepared by first dialyzing all purified antibodies (concentrations ranging from $0.3\,mg\,mL^{-1}$ to $10\,mg\,mL^{-1}$) against $5\,mM\,Na_2HPO_4$, pH 7. The resulting samples were then mixed 1:1 (or 1:4 for 0.5% HPMC containing Pharmalyte) with ampholyte solution. The separation was initiated by applying a constant voltage at $-13\,kV$. The cathode was in the injection side of the capillary. The catholyte consisted of 20 mM NaOH, and the anolyte was 25 mM phosphoric acid. The sample components were detected within 5 minutes by UV absorption at 280 nm. System suitability was assessed by analyzing a mixture of human hemoglobins (A, F, S, and C), or horse heart myoglobin.

Many antibodies were found to precipitate at their point of focusing during cIEF conditions. In these instances, urea was included as a solubilizing agent in the separation medium (Figure 3).

In his discussion, based on the system configuration and on the separation of various protein standards, Lee suggested a titration mechanism involving the capillary wall silanols to explain a perceived inconsistency in his results. According to Lee, since the ampholyte preparation with and without urea used ampholytes pH 5–8, and the effective length of the capillary was 9 cm section (28% of the total) a 0.84

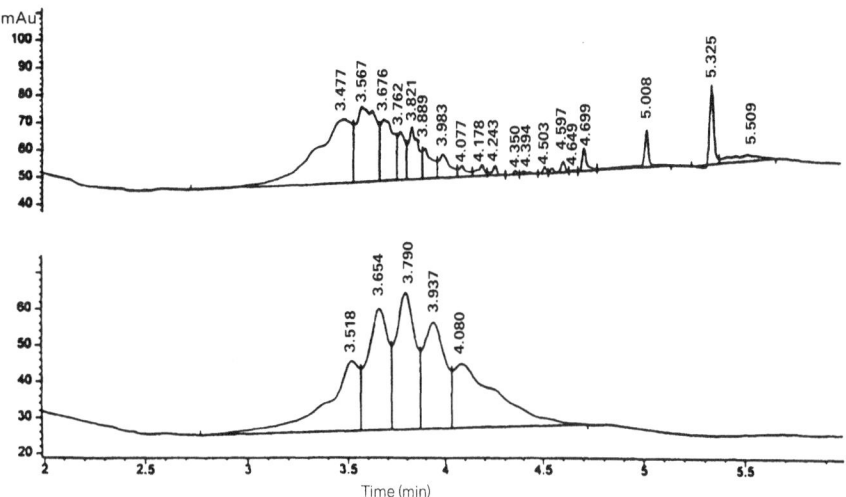

Figure 3. Electropherograms of a monoclonal antibody as analyzed by cIEF in the absence (top) and presence of 7 M urea (bottom). Reproduced from reference 44 with permission.

Table I. Reproducibility of the determination of pI by cIEF in the absence and presence of urea for two monoclonal antibodies.

Condition	Antibody	pI						
		Peak	Eq. 1	Eq.2	Eq.3	Average	S.D.	%RSD
Non-urea	"a"	1	6.656	6.642	6.642	6.641	0.016	0.2
		2	6.577	6.564	6.548	6.563	0.015	0.2
		3	6.499	6.488	6.473	6.487	0.013	0.2
		4	6.416	6.407	6.394	6.406	0.011	0.2
Urea	"f"	1	6.936	6.945	6.942	6.941	0.005	0.1
		2	6.878	6.887	6.884	6.883	0.005	0.1
		3	6.820	6.828	6.825	6.824	0.004	0.1
		4	6.756	6.763	6.761	6.760	0.004	0.1
		5	6.698	6.704	6.702	6.701	0.003	<0.1

Notice that the variation of pI is low for interday assays under both analysis conditions. Each analysis was performed in duplicate. Reproduced from reference 41 with permission.

pH unit range was expected to span the effective length of the capillary. This configuration would allow detection of proteins with pI values between 5.00 and 5.84. However, antibodies and proteins with pI values $\geqslant 5.84$ were still observed. This was also true for the proteins used as pI standards. Since Lee's cIEF procedure was single-step, a stable formation of the pH gradient may never have been achieved. In fact, single step cIEF can be regarded as a form of isotachophoresis with a very large number of spacers (the ampholytes, which are not observed at 280 nm), some of which may act as the leading and trailing ions. Under these circumstances, since the capillary was filled with sample, all components would migrate past the detection point. This migration towards the detector is aided by the residual EOF.

Lee found that his procedure was highly reproducible (Table I) for both migration time and pI. Reproducibility was better than 2% RSD for the migration times, obtained using two sets of reagents

and capillaries on three consecutive days. Normalized pI values were determined to have a day-to-day variability of only 0.01 pH unit.

In an application involving fragments of MAbs, Hagmann et al. [42] used cIEF, electrospray ionization mass spectrometry, and reversed phase HPLC peptide mapping for the characterization of the F(ab')2 fragment of a murine monoclonal antibody. In addition, high performance anion exchange chromatography-pulsed amperometric detection (HPAEC-PAD) was used to identify carbohydrates derived from the glycoprotein.

The authors manufactured a F(ab')2 fragment by pepsin cleavage of a murine IgG3 isolated from the cell-free culture supernatant of a mouse-mouse hybridoma cell line. Pepsin cleaves in the hinge region of the IgG, producing F(ab')2 and Fc' fragments. The analysis of the F(ab')2 fragment by cIEF under non-reducing conditions revealed that the sample was composed of more than one molecular species. Under reducing conditions, the

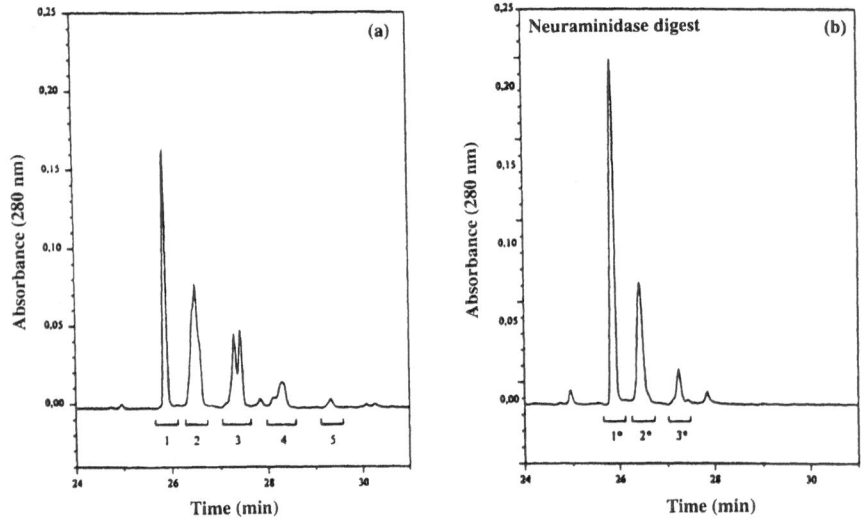

Figure 4. Electropherograms of F(ab')2 by cIEF using chemical mobilization under non-denaturing conditions. (**a**) Capillary IEF pattern of F(ab')2 isoforms. (**b**) Capillary IEF of the same sample after treatment with neuraminidase. The reduced number of peaks after treatment with neuraminidase is due to the removal of sialic acid groups from the polypeptides. Reproduced from reference 42 with permission.

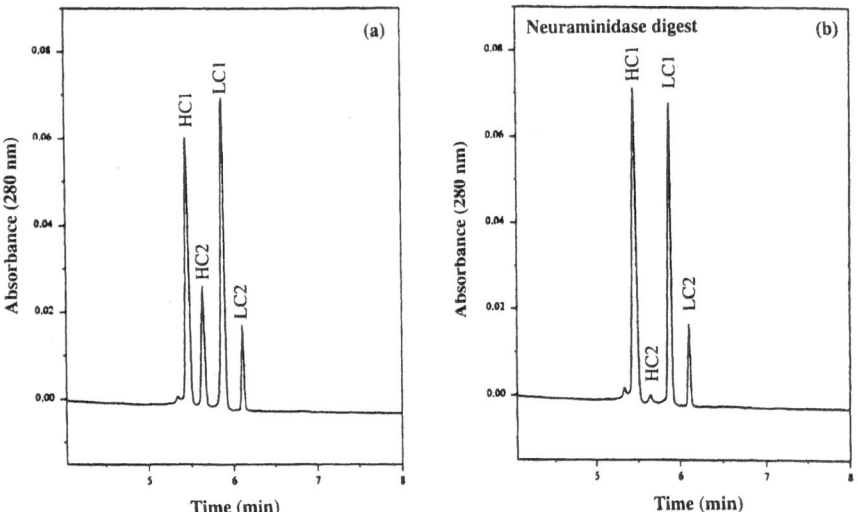

Figure 5. Capillary IEF of F(ab')2 under reducing conditions using EOF mobilization. (a) Isoform pattern of heavy (HC) and light chain (LH). (b) Capillary IEF pattern of the same samples after treatment with neuraminidase. Reproduced from reference 42 with permission.

heavy chain and the light chain were separated in two peaks each. For cIEF, the samples were prepared at concentrations of $300-500\,\mu g\,mL^{-1}$. Each sample was analyzed twice, with and without the addition of pI markers (cytochrome C and myoglobin). The ampholyte mixture consisted of 4% Pharmalyte 3-10, 1% TEMED, and 0.8% methylcellulose. For cIEF under non-denaturing conditions, a coated capillary (μ-Sil DB-1, 27 cm × 50 μm ID, J&W Scientific, Folsom, CA) was used. The capillary was rinsed with 20 mM phosphoric acid between runs. Detection was performed by UV absorption at 280 nm. The catholyte was 20 mM NaOH, and the anolyte 20 mM phosphoric acid. The proteins were focused for 5 minutes at 25 kV. The capillary temperature was maintained at 25 °C. Cathodic chemical mobilization was accomplished by replacing the catholyte with a 20 mM NaOH solution containing 30 mM NaCl. For cIEF under denaturing and reducing condition, a one-step method was used with mobilization. Samples were diluted in 8 M urea and 2% DTT and mixed in a 1:1 ratio with the ampholytes (4% Pharmalyte 3–10, 1.5% TEMED in 8 M urea, 2% DTT). Simultaneous focusing and mobilization took place at 15 kV with reversed polarity in the shorter segment of

the capillary (37 cm × 50 μm ID capillary, eCAP, neutral capillary, Beckman, Fullerton, CA, USA).

When the samples were analyzed by cIEF, several groups of peaks were observed. Analysis by ESI-MS revealed a high mass heterogeneity of the molecule. For mass spectrometry, the sample was dialyzed and its concentration adjusted to $1\,mg\,mL^{-1}$. Digestion with neuraminidase simplified both the cIEF pattern and the mass spectrum (Figure 4). The cIEF of the reduced molecule showed that the sialic acids were located only on the heavy chain of the F(ab')2-fragment (Figure 5). Through comparison of the peak area and the pI of light and heavy chains, the authors concluded that a glycosylation site was located on the heavy chain and a further site for modification was located on the light chain.

By incubation with O-glycosidase a further reduction of the complexity of the mass spectrum was achieved showing 8 different isoforms. By LC-MS peptide mapping these isoforms could be attributed to the heterogeneity of the pepsin cleavage site in the hinge region of the antibody. Although pepsin is considered to cleave predominantly at aromatic and hydrophobic residues such phe and leu, in reality it has a broad spectrum of cleavage sites.

LC-MS was performed using a C18 capillary column (250 × 1 mm) with an acetronitrile gradient from 5 to 56% over 170 min. The mobile phases contained 0.1% TFA. To determine the amino acid residue modified by glycosylation, peptide analysis by LC/MS was performed after the antibody was digested with the endoproteinase LysC (25 mM Tris, 1 mM EDTA, pH 8.5 for 18 hrs at 37 °C).

The sugars of the O-linked carbohydrate chain were identified by HPAEC-PAD as galactosyl-N-acetyl-galactosamine (GalNAcGal) with terminal N-glycolylneuraminic acid. The glycosylation site was identified by peptide mapping and amino acid sequence analysis as Ser_{222}. Murine IgGs carry one N-glycosylation site on the CH_2 domain of the heavy chain. However, N- as well as O-glycosylation have also been reported to occur in the Fab part of IgGs. Two of the peaks had altered masses after enzymatic removal of the sugar moieties. For the identification of the glycosylated amino acid residues the two glycopeptide containing peaks were sequenced. In the sequencing cycle where Ser_{222} was expected, no phe-

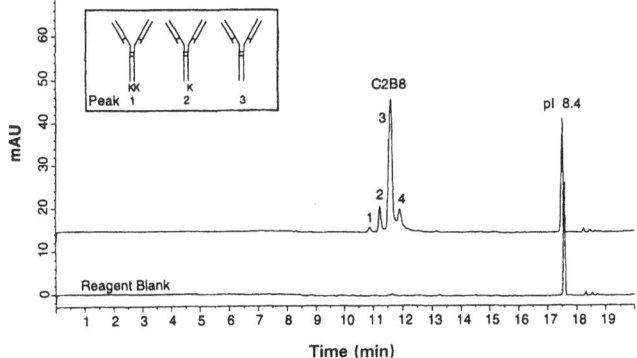
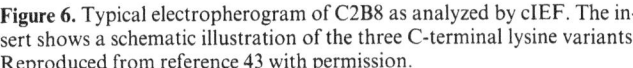

Figure 6. Typical electropherogram of C2B8 as analyzed by cIEF. The insert shows a schematic illustration of the three C-terminal lysine variants. Reproduced from reference 43 with permission.

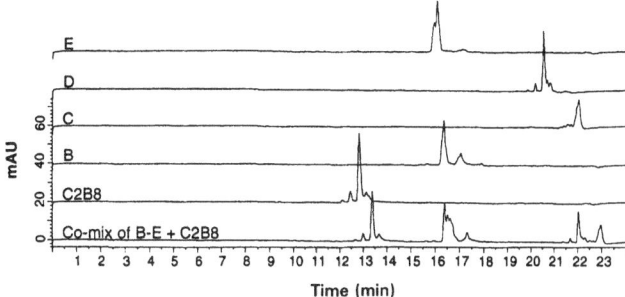

Figure 7. Separation pattern of C2B8 compared with other monoclonal antibody preparations. Reproduced from reference 43 with permission.

nylthiohydantoin (PTH)-amino acid was found. The results showed that only Ser_{222} was glycosylated.

During ESI-MS, it was observed that after treatment with neuraminidase a mass shift of about 307 was observed for some of the peaks, which correlated with the removal of one N-glycolyl neuraminic acid (NGNA) residue. After treatment with O-glycosidase, an additional mass shift of 365 occurred, which fitted the known substrate specificity of O-glycosidase for the disaccharide galactosyl-N-acetyl-galactosamine (Gal-NAcGal).The identity of the sugars was further supported by matching their HPAEC-PAD retention times with those of oligosaccharide standards. Carbohydrates were separated in a Carbo-Pac PA-100 column (Dionex, Sunnyvale, CA, USA) at a flow rate of 1 mL per minute using a biphasic acetate gradient in 100 mM NaOH (2–10 minutes from 0 to 20 mM sodium acetate, 10–25 minutes from 20 to 250 mM sodium acetate). Detection was performed using an amperometric detector. A very small amount of N-acetylneuraminic acid (NANA) was also found.

The two isoforms of the light chain were not influenced by the neuraminidase treatment. Thus, glycosylation can be excluded as a source of the observed modification of the light chain. Since deamidation occurs preferentially at an asparaginyl residue with glycine or serine on the C-terminal side, and the sequence of the light chain includes an asp^{125}-gly^{126}, deamidation is very likely the cause for the observed charge heterogeneity of the light chain. Protein deamidation leads to an acidic shift of the polypeptide's pI due to the generation of an additional carboxylic group. For F(ab')$_2$ this assumption was supported by ESI-MS of the reduced sam-

ples. According to the authors, only deamidation can cause an acidic shift of the pI together with a mass difference of < 2. After sugar removal, the C-terminal heterogeneity was proven using the molecular masses of LC-MS peptide maps.

Hunt et al. [43] validated a capillary isoelectric focusing method for determination of the identity and charge heterogeneity of the recombinant monoclonal antibody C2B8. C2B8 is a mouse/human chimeric monoclonal antibody directed to the human CD20 antigen. The variable region of the light and heavy chains is murine. One conserved Asn-linked glycosylation site is found within the constant region of each heavy chain. There is also possible charge heterogeneity resulting from C-terminal modification and deamidation.

The assay was validated in accordance with ICH guidelines in order to demonstrate that it was suitable for its intended purpose as a lot release test for the bulk and final product. The criteria for a successful method validation are understandably stringent and are outlined in the guidelines of the "International Conference on Harmonization (ICH) of Technical Requirements for Registration of Pharmaceuticals for Human Use". The application of the validation process in accordance with the ICH guidelines is imperative for the scientist seeking to validate a method for Biological License Application (BLA) and Marketing Authorization Application (MAA) submissions.

During the feasibility phase of method development, a first pass at "validation" may include intra and inter-day precision, linearity, and specificity. But for a certificate of analysis test, additional parameters must be evaluated such as identity, purity, main peak assays, and recovery.

In the performance of a method there are usually several critical components which must be evaluated to assess the quality and reproducibility of manufacturing. To this regard, Hunt et al. also assessed the performance of various lots of capillaries, alternate capillaries, and alternate instruments.

As a result of the validation process, the assay was found to be linear over the concentration range of 2–356 µg mL^{-1}. Recovery was determined using ^{125}I-labeled C2B8 at the targeted sample concentration of 125 µg mL^{-1}. As control (100% recovery), the sample was injected, focused, and then expelled from the capillary with water. To determine recovery, the sample was injected, focused, mobilized (mobilization was stopped 30 sec after the last peak had passed the detection point), and finally expelled of the capillary with water. Comparison of the control and experimental samples resulted in a mean recovery of 99%.

Specificity was evaluated by showing resolution of the main components from excipients, degradation products (Figure 6), and other monoclonal antibodies (Figure 7). Repeatability and intermediate precision were assessed for the four major peaks. Observed % RSD values for migration time, peak area, and peak area percent ranged from 0.9–4.4% (Table II). The results of this validation demonstrated that the cIEF assay for the determination of identity and charge distribution of C2B8 was accurate, precise, linear, and highly specific. The authors demonstrated that the assay was rapid and suitably rugged. Examples demonstrating method robustness are presented in Figure 8, and its use as a stability indicating assay is shown in Figures 9 and 10.

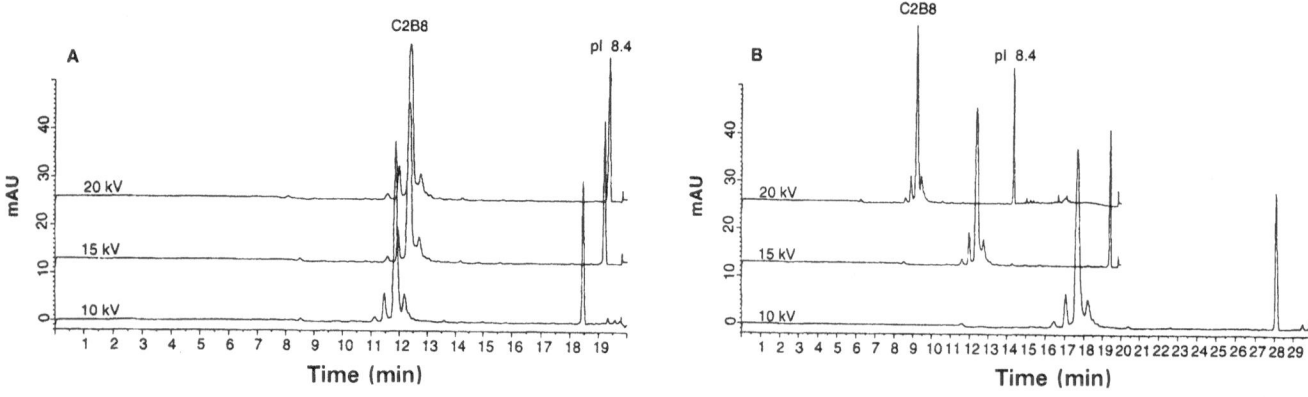

Figure 8. Analysis of a single sample of C2B8 by cIEF using different focusing (**a**) and mobilization (**b**) voltages. Reproduced from reference 43 with permission.

Table II. Repeatability and intermediate precision in the analysis of C2B8 by cIEF.

Analysis	Value	Peak			
		1	2	3	4
Migration time (min)	Mean	11.7	12.0	12.4	12.7
Repeatability[a, c]	RSD (%)	1.4	1.2	1.2	1.1
Intermediate[b, d]	Mean	11.1	11.5	11.8	12.2
	RSD (%)	2.0	1.8	1.9	1.8
Peak area (μVs)	Mean	8.5	42.0	289.0	73.2
Repeatability[a, c]	RSD (%)	1.9	1.4	0.5	2.1
Intermediate[b, d]	Mean	9.2	41.2	297.5	79.1
	RSD (%)	4.2	1.4	1.9	4.4
Area (%)	Mean	2.1	10.2	70.2	17.8
Repeatability[a, c]	RSD (%)	1.6	0.7	0.5	0.7
Intermediate[b, d]	Mean	2.2	9.7	69.7	19.2
	RSD (%)	3.7	1.1	0.9	2.1

[a] A single preparation of C2B8 was analyzed. [b] Two analysts in different laboratories independently analyzed a sample on three separate days using fresh samples prepared each day. [c] values represent the mean of six replicate injections. [d] each value represents the mean of six injections (duplicate injections on three separate days). Reproduced from reference 43 with permission.

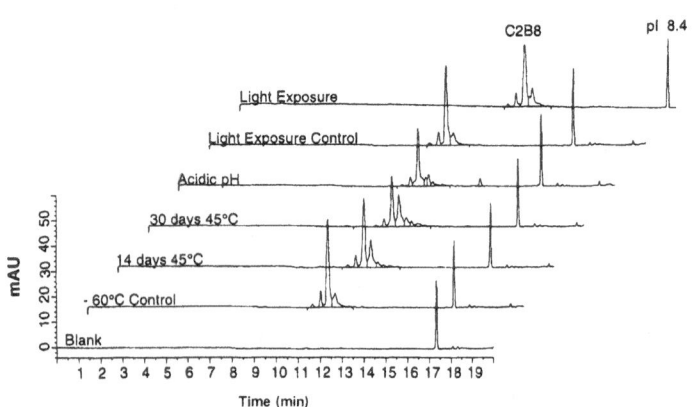

Figure 9. Capillary IEF analysis of a C2B8 sample subjected to various modes of degradation. These results clearly show the stability indicating capabilities of cIEF for these modes of degradation. Reproduced from reference 43 with permission.

6.7 Capillary Isoelectric Focusing of Hemoglobin and Hemoglobin Variants

Hemoglobins are the major component of red blood cells, which transport oxygen to the body tissues and facilitate the return transport of carbon dioxide. Hemoglobins are very soluble in water, and they are present in erythrocytes at a concentration that can exceed 300 mg mL^{-1}. Hemoglobins are formed by four globin chains: two α-globin, and two β-, δ, or γ-globin chains, each bound to a heme group.

Mammalian hemoglobins have molecular weights of approximately 64.5 kDa. Most hemoglobins and Hb variants have pIs ranging from approx. 6.5 to 8.0. Capillary IEF can easily resolve adult Hb (Hb A) from normal variants such as fetal Hb (Hb F) which is found in blood up to the age of six months after birth (contributing from 60–80% to the total Hb). At this time, Hb F is gradually and largely replaced by Hb A. Normal adult blood contains less than 1% of Hb F, and between 2 and 3% Hb A$_2$. Capillary IEF can also be used to distinguish abnormal Hb species associated with a variety of blood disorders. Disorders arise from defective genes that code for altered Hb chain sequences, and abnormal levels of individual globin chains produce novel Hb tetramers, which are also characteristic of the disorder. A high number of point mutations (> 600) in Hb chains has been estimated.

Since many congenital and acquired disorders are associated with abnormal levels of normal globin chains, or production of altered globins, analysis of hemoglobin (Hb) variants is of major importance in clinical diagnostics. Various electrophoretic approaches, including cellulose acetate electrophoresis at alkaline pH, citrate agar electrophoresis at acidic pH, and isoelectric focusing (IEF) in polyacrylamide or agarose gels as well as HPLC, immunological assays, structure analysis, and genotypic methods have been used to investigate hemoglobinopathies [45]. HPLC is now considered to be a sensitive, specific, and reproducible alternative to electrophoresis and its use has been significantly expanded, especially with the development of rapid and well-resolving methods. In the field of newborn screening programs, IEF emerges as a remarkable resolving method that allows unequivocal identification of a large number of Hb variants.

Capillary IEF is a rapid (< 15 minutes), precise (coefficient of variance < 2%), sensitive, and high resolution automated method for the routine quantitative analysis of hemoglobin (Hb) variants using a minimal amount (< 10 µl) of blood. Capillary IEF is comparable to high performance liquid chromatography (HPLC), but with higher resolution and much lower cost of consumables. Sample preparation is also simplified, because unlike most other Hb assay methods, minor variants of hemoglobin are routinely measured by cIEF along with the major hemoglobin constituents [46].

Although the number of mutations detected in globin chains is large, a significant number of these mutations are silent. Other mutations are very rare, and more than one variant may show in the same individual. For example Warrier et al. [47] described a rare association of Hb C and E in an infant whose blood showed abnormal behavior in a neonatal screening program. Family studies showed that the mother's blood contained Hb C and Hb A, the father's Hb A and Hb E, and a sibling's Hb A and Hb C. Using standard electrophoretic techniques, Hb E was not resolved.

Many cIEF configurations have been used to analyze hemoglobins. Due to the large number of known Hb variants, resolution is critical for any method developed for their analysis. Both single-step and two-step cIEF procedures have been used, and various capillaries, anolyte and catholyte compositions, and ampholyte mixes have been explored. As expected, the largest effect on resolution was achieved by manipulating the ampholyte composition. Most of the publications reviewed here used ampholytes covering the pH range of 3–10 with 6–8 ampholytes added in various ratios. In most cases, the ampholyte-sample solution contained polymers in order to increase the viscosity of the solution, and in some cases to reduce electro-osmosis, and non-specific interaction of the sample with the capillary wall.

Hempe et al. [48] revised their method, which has been in use since approximately 1995 for the analysis of hemoglobin variants by cIEF. In their revised method, the authors use red blood cells (RBCs) instead of whole blood. To improve resolution, mixed ampholytes were used (pH ranges 6–8 and 3–10 mixed at a 9:1 ratio, 2% solids), and the viscosity of the solu-

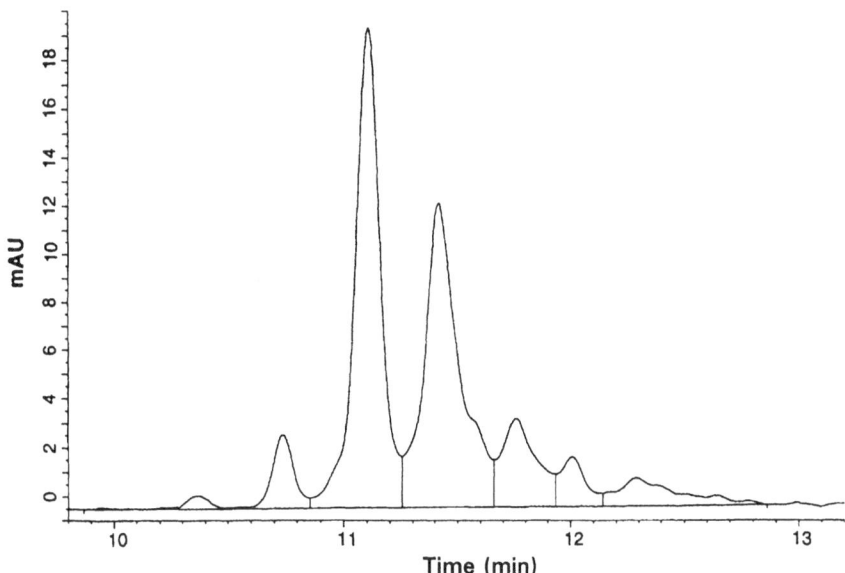

Figure 10. Expanded view of a C2B8 sample held at 45 °C for 30 days. Reproduced from reference 43 with permission.

tion was controlled by adding 0.375% methylcellulose. A DB-1 (J&W Scientific, Folson, CA, USA) capillary was used, conditioned with a methanol rinse between runs. Detection was performed by UV absorption at 415 nm. Identification of Hb variants was based on pI determined by linear regression of the migration time against the pI of known Hb variants in a control sample. Their results showed that cIEF is rapid (15 min per sample) and sensitive for the detection of Hb variants present in low concentrations (e. g. Hb F, Hb A_2, and Hb A_2'). Hemoglobins C and E (pI 7.44 and 7.41), and S and D (pI 7.21 and 7.18) were easily resolved from each other, demonstrating the high resolving power of cIEF. Hempe et al. [49] also analyzed Hb A_{1C}, which is widely used as an indicator of long-term glycemic control in diabetes. The focused protein bands were mobilized by applying low pressure at the end of focusing. Stable Hb A_{1C} was measured in clinical samples after removing the labile fraction by incubating 0.1 mL of RBC in 1 mL of saline solution at 37 °C for 6 hours. Longer incubation or hemolysis of the RBC at low pH caused the formation of methemoglobin oxidation products and glutathione adducts that interfere with quantitation. The cIEF reference range determined in 124 non-diabetic subjects was 5.3% to 8.4% (mean ± 2 SD). Interference by carbamylated Hb and high levels of acetylated Hb F was observed.

Craver et al. [46] found a clear distinction in Hb A_2 levels in healthy persons and patients with sickle cell trait or sickle cell anemia compared with those in patients with β-thalassemia. Elevated Hb A_2 is an important diagnostic indicator for β-thalassemia. Although it is well known that Hb A_2 levels are also higher in sickle cell trait or sickle cell anemia without the features of β-thalassemia, reference intervals for Hb A_2 in these conditions are not generally used because Hb A_2 levels are not routinely measured with standard analytical procedures. The slightly higher Hb A_2 level normally present with sickle cell trait and sickle cell anemia may lead to erroneous diagnosis of these diseases with β-thalassemia.

All separations were performed with capillary cooling at 20 °C using a 50 µm × 27 cm DB-1 (0.05 µm coating thickness, J&W Scientific, Folsom, CA) capillary. The ampholytes (2% v/v of a 10:1 mixture of Pharmalyte 6–8 and 3–10 [Pharmacia Biotech Products, Piscataway, NJ, USA]) contained 0.375% w/v methyl cellulose (1500 cps at 20 g L⁻¹). The anolyte was 100 mM phosphoric acid in methyl cellulose solution. The catholyte was 20 mM NaOH in deionized, distilled water. Focusing was carried out for 5 minutes at 30 kV. Pressure was used for mobilization. The whole blood was hemolyzed in 5 mM EDTA + 10 mM KCN. Unlike most other clinical Hb analysis methods, Hb A_2 and other minor variants were routinely measured by cIEF along with the major hemoglobin constituents.

Normal reference intervals for Hb A_2 levels had not yet been established for the cIEF method. From 862 quantitative cIEF typings, the authors established mean values and reference intervals based on those in 412 healthy persons classified into age groups. Mean values and reference intervals for Hb A_2 were then determined for patients one year of age or older with sickle cell trait, sickle cell anemia, and β-thalassemia, and these were compared with each other and with normal values. Simultaneously measuring major and minor hemoglobin variants by cIEF, the authors obtained Hb A_2 intervals in healthy volunteers ($n = 412$) (reference value) and patients with Hb S or β-thalassemia. The normal Hb A_2 reference interval was classified into three age groups:

5 months or younger	$1.2\% \pm 1.5\%$
6 months to 1 year	$2.2\% \pm 0.9\%$
1 year or older	$2.4\% \pm 0.9\%$

No difference was noted between the one-to two-year old patients, and those patients older than 2 years, thus they were all grouped as "one year or older".

These intervals were comparable to those used with other methods. Patients one year of age or older with Hb S had significantly higher Hb A_2 levels ($P < 0.5$):

Sickle cell trait	$2.9\% \pm 0.09\%$
Sickle cell anemia	$2.8\% \pm 1.0\%$

Although reference Hb A_2 intervals overlapped those in patients with Hb S, no overlap in Hb A_2 levels was noted between these groups and patients with β-thalassemia:

β-thalassemia	4.3% to 7.5%

The higher than normal Hb A_2 interval in patients with Hb S must be considered before a diagnosis of sickle cell trait or sickle cell disease with β-thalassemia is made. Iron deficiency can decrease Hb A_2 levels in β-thalassemia to within the normal range. Mean values for sickle cell trait, sickle cell anemia, and β-thalassemia were significantly different from normal values, and values for sickle cell trait and sickle cell anemia were different from those of β-thalassemia (all $P < 0.05$). Mean values for sickle cell trait and sickle cell anemia were not significantly different.

Mohammad et al. [50] proposed the use of a "migration index" to overcome the variability of migration times observed in cIEF performed in fused silica capillaries. The authors introduced two synthetic pI (6.6 and 7.7) markers (Bio-Rad Laboratories, Hercules, CA, USA) bracketing the pI gradient formed by the ampholytes used (6.6 to 7.7).

Migration index =
$(T_{Hb} - T_{7.7}) \div (T_{6.6} - T_{Hb})$

Where T_{Hb}, $T_{7.7}$, $T_{6.6}$ are the migration times of the hemoglobin variant, the pI 7.7, and the pI 6.6 markers, respectively. Using this method, the CV% dropped from about 15% for migration time to less than 5% for migration index. Capillary IEF was performed in a 50 μm × 57 cm fused silica capillary. The catholyte was 20 mM NaOH and the anolyte was 100 mM phosphoric acid. The ampholyte concentration was $50 \, \text{ml L}^{-1}$ (the authors did not specify the pH range) containing $3 \, \text{g L}^{-1}$ methyl cellulose. The capillary was first filled with ampholyte solution, then a sample plug was injected. Focusing was carried out at 30 kV for 5 minutes, and the separated Hb variants were mobilized past the detector window by applying low pressure (80 mbar) under an applied voltage of 30 kV.

Because of the existence of considerable EOF, this method was unable to consistently resolve hemoglobin C from Hb A_2 and Hb E.

Mario et al. [45] developed two assays for the complete analysis of Hb variants: a high-performance cation-exchange liquid chromatography (HPLC) assay using a polyaspartic acid weak cation-exchanger, and two-step cIEF on a neutral-coated capillary in a narrow pH gradient. The HPLC method used a Poly Cat A column (Touzart & Matignon, Les Ulis, France) with salt gradient elution. Capillary IEF separations were performed on a 50 μm × 37 cm eCAP neutral capillary (Beckman Coulter, Fullerton, CA, USA). The ampholytes were a mixture of Pharmalyte 6–8 and Pharmalyte 7–9 (3:1 v/v), at a final concentration of $20 \, \text{ml L}^{-1}$ in a $4 \, \text{g L}^{-1}$ methylcellulose solution (1500 cp at 2%). Before each run, the capillary was filled with sample and then backflushed with the catholyte to limit the focusing area to the capillary section anodal to the detection window. The sample was focused for 3 minutes at 30 kV, and then mobilized by pressure while maintaining a voltage of 25 kV.

The resolution was satisfactory for both methods. The resolution for cIEF was estimated to be 0.02 pH units. Both techniques allowed separation of normal and common abnormal Hbs, i. e. Hb A, Hb A_2, Hb S, Hb C, Hb F, and Hb E. With HPLC, Hb A_2 and E were completely separated from Hb C (which carries a glutamic acid to lysine transition at position 6 of β-globin), but C-Harlem (carrying the C mutation and an aspartic acid to asparagine transition at position 73 of β-globin) co-eluted with Hb C, and Hb E co-eluted with Hb A_2. HPLC was unique in distinguishing homozygous Hb C disease from double heterozygous inheritance of Hb C/$β_0$-thalassemia, in which Hb A_2 is increased. With cIEF, Hb A_2, Hb E, and Hb C-Harlem were partially separated from Hb C, but Hb E co-migrated with Hb A_2.

Quantitative precision of Hb A_2 was evaluated at two concentrations, and RSD values were 7.5% and 2.9% for the low and high levels, respectively. The RSD values for Hb F and Hb S were 1.4% and 5.4%, respectively. A low Hb A_2 level obtained by cIEF may be related to the separation of glycated Hb A_2. Quantitative data obtained for Hb A_2, F, and S were highly correlated between the two different assays.

Both cIEF and HPLC allowed the diagnosis of β-thalassemia by an accurate quantification of both Hb F and Hb A_2 in a single run, without overlap between Hb A_2 values for normal individuals and for β-thalassemia patients. For this differentiation, both assays were more convenient than classical methods, which need two experiments, i. e. alkali denaturation for Hb F, and microcolumn ion-exchange for Hb A_2. Intra-day reproducibility was calculated from the retention times of Hb A obtained in 20 successive runs. Inter-day reproducibility was estimated from the retention times obtained in 10 different days with the use of different reagent preparations. The intra-day CV was 1.2% for HPLC, and 10% for cIEF. The inter-day CV for HPLC was 3.3% and CV 4.9% for cIEF.

In another publication, Mario et al. [51] described their development of two cIEF methods for the analysis of hemoglobin variants: a one-step and a two-step cIEF in narrow pH gradients that could be used for the analysis and quantitation of abnormal hemoglobins. The separations were performed in a 37 cm × 50 μm neutral coated capillary. The one-step method used residual electroosmosis to transport the focused proteins past the detector point. This procedure was performed in reversed polarity, with the catholyte at the injection end of the capillary, and the anolyte at the end closest to the detector. The capillary was filled with the sample mixed with 5% ampholyte pH 6–8 solution in 0.2% HPMC (4000 cp at 2%) and supplemented with 0.75% TEMED to prevent sample focusing past the detector window. Application of the electric field (7 kV) simultaneously focused and mobilized the hemoglobins.

The two-step procedure was performed with the catholyte closest to the detector. The focused samples were mobilized by applying low pressure at the end of focusing. The capillary was filled with the 2% ampholyte solution (75:25 pH 6–8:pH 7–9) in 0.4% HPMC (1500 cp at 2%), followed by a NaOH backflush to limit the focusing area to the effective length of the capillary. The sample was focused for 3 minutes at 30 kV and then mobilized by low pressure maintaining a high voltage (25 kV).Detection was by absorption at 415 nm. Because of the selective absorption of the heme at 415 nm, all bands detected were attributed to hemoglobins. By using the slope of the gradient (as determined by the migration time of hemoglobin standards), the authors estimated a resolution of 0.1 pH unit for the one-step cIEF, and 0.02 pH unit for two-step cIEF. The single step method allowed the separation of the Hb A, S, F, and C, but Hb A_2 could not be quantitated. The two-step cIEF method allowed separation of some Hb variants that co-migrate with Hb A2 in alkaline electrophoresis: Hb E, and Hb C-Harlem. Retention time reprobucibility showed a CV < 5%.

Although both methods were performed in the same type of capillary, the authors believed that in the one-step cIEF the samples were mobilized by the residual electroosmotic flow. According to their configuration, it is more likely that they were observing focusing peaks.

6.8 Capillary Isoelectric Focusing of Erythropoietin

Capillary IEF of erythropoietin (EPO) glycoforms was performed by Cifuentes et al. [52] and compared to flat-bed IEF and CZE. EPO regulates erythropoiesis, the process that controls the production of erythrocytes in mammals. The recombinant human (rhEPO) form of this molecule is of great economic interest, because it can be used to correct various sources of anemia, and as a transfusion substitute. Re-

combinant human EPO produced in Chinese hamster ovary (CHO) cells is a glycoprotein with a molecular mass of 30.4 kDa, and a carbohydrate content of 39.5%, which shows a protein conformation apparently identical with the natural product isolated from human urine. Since glycosylation of recombinant proteins depends, among other factors, on the host cell line used for its production, the carbohydrate structure of natural and recombinant EPO may be different. It is known that the sialic acid content of EPO plays an important role in its biological activity. The change in pI due to the negative charge contributed by each sialic acid group can be exploited in electrophoresis to resolve the various glycoforms. Prior to analysis, rhEPO was dissolved in water. The low molecular excipients were eliminated by passage through a 10 kD cutoff ultrafiltration cartridge and the retentate was washed three times and reconstituted in water. Laboratory-made polyacrylamide-coated capillaries (27 cm × 50 μm) were used for cIEF analysis. Ampholytes 3-10, 2.5-5, and a mix of the two ranges were used. The best results were obtained with a 1:2 mixture of ampholytes 3-10 and 2.5-5 respectively. Only the mixtures of ampholytes containing 7 M urea produced acceptable separation of the rhEPO isoforms.

When the samples were completely desalted, a very low current was produced at the beginning of focusing, but the EPO glycoforms could not be separated. When the sample was mixed with an equal volume of the second wash flow-through (which still contained some of the excipients) the results were greatly improved. The authors also attempted to "desalt" the samples by ramping of the voltage, but they observed that the current was too high even at very low voltages, and there was almost no separation of the EPO glycoforms. In our experience, this "desalting" practice leads to pH gradient compression, and thus, to loss of resolution.

Resolution of rhEPO glycoforms was best achieved in uncoated capillaries by zone electrophoresis in the presence of 0.025 M putrescine. The analysis time for this technique was about 30 minutes (as opposed to 15 min for cIEF).

6.9 Novel Applications of cIEF

6.9.1 Use of cIEF as a Detection System for Hybridization Probes

Use of cIEF as a detection method for fluorescently-labeled nucleic acid hybridization probes has been demonstrated by Cruickshank et al. [53]. Peptides containing cysteine and lysine residues were converted to fluorescent species by linkage of Rhodamine Green via the lysine residue. A group of eleven different labeled peptides were shown to exhibit unique pI values in cIEF using electrophoretic mobilization in a neutral coated capillary. For use as hybridization probes, 5'-amino functionalized oligonucleotides were converted to S-pyridyldithiopropionyl derivatives. Each oligonucleotide derivative was then conjugated to a specific fluorescently-labeled peptide via disulfide linkage and used in a hybridization reaction. Following hybridization, the signal peptides were released by treatment with cysteine and assayed by cIEF with laser-induced fluorescence detection. The authors proposed that the unique pI of each labeled peptide and the high resolution of cIEF allow this system to be used as a multiple probe detection system. Satisfactory accuracy and reproducibility of the cIEF assay relied upon the use of fluorescently-labeled peptide sequences as internal standards which flanked the released signal peptides.

6.9.2 Affinity cIEF

Capillary IEF using electrophoretic mobilization in a neutral coated capillary was used to study the binding of biotin to actinavidin by Okun [54]. The unbound protein exhibited a large number of peaks which were thought to represent different conformational states. Binding of biotin or biotinylated oligonucleotide resulted in the appearance of fewer components with lower apparent pI values. At high biotin-oligonucleotide concentrations, actinavidin isoforms disappeared and a single species was observed. The authors interpreted the reduction in microheterogeneity as an indication that ligand binding stabilizes the structure of actinavidin, yielding fewer conformational isoforms.

Rhigetti et al. [55] have described the combined use of CZE and cIEF to study protein-protein interactions. The complexation of hemoglobin (Hb) with haptoglobin was investigated by focusing hemoglobin (pI 7.0) within the capillary in a pH 6–8 ampholyte gradient. Haptoglobin was introduced at the cathode and swept electrophoretically through the focused Hb zone. Hemoglobin-haptoglobin complexes have pI values between 5.0–5.50, outside the range of the gradient. The complexes were detected at 416 nm as they migrated electrophoretically past the detection window. As an additional check on binding stoichiometry, the pH gradient could be disrupted and the remaining hemoglobin mobilized past the detector.

6.10 References

[1] Wehr, T.; Zhu, M.; Rodriguez-Diaz, R. "Capillary Isoelectric Focusing," in *Methods in Enzymology* Vol. *270*, Karger, B.L.; Hancock, W.S. (eds.), Academic Press, San Diego, **1996**, pp. 358–374.

[2] Rodriguez-Diaz, R.; Wehr, T.; Zhu, M.; Levi, V. "Capillary Isoelectric Focusing," in *Handbook of Capillary Electrophoresis*, 2nd Edition, Landers, J.P. (ed.), CRC Press, Boca Raton, **1996**, pp. 101–153.

[3] Wehr, T.; Rodriguez-Daz, R.; Zhu, M. Capillary Electrophoresis of Proteins (*J. Chromatogr. Library,* Vol. *80*) Marcel Dekker, New York, **1998**, pp. 131–233.

[4] Hjertén, S. "Isoelectric Focusing in Capillaries," in *Capillary Electrophoresis: Theory and Practice*, Grossman, P.D.; Colburn, J.C. (eds.) Academic Press, San Diego, CA, **1992**, Chapter 7.

[5] Zhu, M.; Rodriguez, R.; Wehr, T. *J. Chromatogr.* **1991**, *559*, 479.

[6] Righetti, P.G.; Bossi, A.; Gelfi, C. *J. Cap. Elec.* **1997**, *4*, 47.

[7] Mazzeo, J.R.; Krull, I.S. *Anal. Chem.* **1991**, *63*, 2852.

[8] Mazzeo, J.R.; Krull, I.S. *J. Chromatogr.* **1992**, *606*, 291.

[9] Tang, Q.; Lee, C.S. *J. Chromatogr. A* **1997**, *781*, 113.

[10] Thormann, W.; Caslavska, J.; Molteni, S.; Chmelik, J. *J. Chromatogr.* **1992**, **589**, 321.

[11] Whynot, D.M.; Hartwick, R.A.; Bane, S. *J. Chromatogr. A* **1997**, *767*, 231.

[12] Hjertén, S. *J. Chromatogr.* **1985**, *347*, 191.

[13] Kilár, F.; Végvári, Á.; Mód, A. *J. Chromatogr. A* **1998**, *813*, 349.

[14] Hjertén, S.; Zhu, M. *J. Chromatogr.* **1985**, *346*, 265.

[15] Rodriguez, R.; Zhu, M.; Wehr, T. *J. Chromatogr. A* **1997**, *772*, 145.

[16] Manabe, T.; Miyamoto, H.; Iwasaki, A. *Electrophoresis* **1997**, *18*, 92.

[17] Minárik, M.; Groiss, F.; Gas, B.; Blaas, D.; Kenndler, E. *J. Chromatogr. A* **1996**, *738*, 123.

[18] Slais, K.; Friedl, Z. *J. Chromatogr. A* **1994**, *661*, 249.

[19] Kobayashi, A.H.; Aoki, M.; Suzuki, M.; Yanagisawa, A.; Arai, E. *J. Chromatogr. A* **1997**, *772*, 137–144.

[20] Liao, J.-L.; Zhang, R. *J. Chromatogr. A* **1994**, *684*, 143.

[21] Clarke, N.J.; Tomlinson, A.J.; Schomburg, G.; Naylor, S. *Anal. Chem.* **1997**, *69*, 2786.

[22] Conti, M.; Galassi, M.; Bossi, A.; Righetti, P.G. *J. Chromatogr.* **1997**, *757*, 237.

[23] Hofmann, O.; Che, D.; Cruickshank, K.A.; Müller, U.R. *Anal. Chem.* **1999**, *71*, 678.

[24] Rossier, J.S.; Schwarz, A.; Reymond, F.; Ferrigno, R.; Rianchi, F.; Girault, H.H. *Electrophoresis* **1999**, *20*, 727.

[25] Wu, J.; Pawlyszyn, J. *Anal. Chem.* **1995**, *67*, 2010.

[26] Fang, X.; Tragas, C.; Wu, J.; Mao, Q.; Pawliszyn, J. *Electrophoresis* **1998**, *19*, 2290.

[27] Wu, J.; Watson, A. *J. Chromatogr. B* **1998**, *714*, 113.

[28] Wu, J.; Li, S.-C.; Watson, A. *J. Chromatogr. A* **1998**, *817*, 163.

[29] O'Farrel, P.H. *J. Biol. Chem.* **1974**, *250*, 4007.

[30] Banks, J.F. *Electrophoresis* **1997**, *18*, 2255.

[31] Tang, Q.; Harrata, A.K.; Lee, C.S. *Anal. Chem.* **1995**, *67*, 3515.

[32] Yang, L.; Tang, Q.; Harrata, A.K.; Lee, C.S. *Anal. Biochem.* **1996**, *243*, 140.

[33] Wei, J.; Yang, L.; Harrata, A.K.; Lee, C.S. *Electrophoresis* **1998**, *19*, 2356.

[34] Tang, Q.; Harrata, A.K.; Lee, C.S. *Anal. Chem.* **1997**, *69*, 3177.

[35] Wei, J.; Lee, C.S.; Lazar, I.M.; Lee, M.L. *J. Microcol. Sep.* **1999**, *11*, 193.

[36] Lamoree, M.H.; Tjaden, U.R.; van der Greef, J. *J. Chromatogr. A* **1997**, *777*, 31.

[37] Lamoree, M.H.; van der Hoeven, R.A.M.; Tjaden, U.R.; van der Greef, J. *J. Mass Spectrom.* **1998**, *33*, 453.

[38] Yang, L.; Lee, C.S.; Hofstadler, S.A.; Smith, R.D. *Anal. Chem.* **1998**, *70*, 4945.

[39] Severs, J.C.; Hofstadler, S.A.; Zhao, Z.; Senh, R.T.; Smith, R.D. *Electrophoresis* **1996**, *17*, 1808.

[40] Jensen, P.K.; Pasa-Tolic, L.; Anderson, G.A.; Horner, J.A.; Lipton, M.S.; Bruce, J.E.; Smith, R.D. *Anal. Chem.* **1999**, *71*, 2076.

[41] Michalke, B.; Schramel, P. *J. Chromatogr. A* **1998**, *807*, 71.

[42] Hagmann, M.L.; Kionka, C.; Schreiner, M.; Schwer, C. *J. Chromatogr. A* **1998**, *816*, 49.

[43] Hunt, G.; Hotaling, T.; Chen, A.B. *J. Chromatogr. A* **1998**, *800*, 355.

[44] Lee, H.G. *J. Chromatogr. A* **1997**, *790*, 215.

[45] Mario, N.; Baudin, B.; Aussel, C.; Giboudeau, J. *Clin. Chem.* **1997**, *43*, 2137.

[46] Craver, R.D.; Abermanis, J.G.; Warrier, R.P.; Ode, D.L.; Hempe, J.M. *American J. Clin. Pathology* **1997**, *107*, 88.

[47] Warrier, R.P.; Nocerino, A.; Hempe, J.; Craver, R. *Pediatric Research* **1999**, *45*, PP. 153A.

[48] Hempe, J.M.; Granger, J.N.; Warrier, R.P.; Ode, D.L.; Craver, R.D. *Clin. Chem.* **1997**, *43*, S155.

[49] Hempe, J.M.; Granger, J.N.; Vargas, A.; Craver, R.D. *Clin. Chem.* **1997**, *43*, S252.

[50] Mohammad, A.A.; Okorodudu, A.O.; Bissell, M.G.; Dow, P.; Reger, G.; Meier, A.; Guodagno, P.; Petersen, J.R. *Clin. Chem.* **1997**, *43*, 1798.

[51] Mario, N.; Baudin, B.; Giboudeau, J.J. *J. Chromatogr. B* **1998**, *706* 123.

[52] Cifuentes, A.; Moreno-Arribas, M.V.; de Frutos, M.; Diez-Masa, J.C. *J. Chromatogr. A* **1999**, *830* 453.

[53] Cruickshank, K.A.; Olvera, J.; Muller, U.R. *J. Chromatogr. A* **1998**, *817*, 41.

[54] Okun, V.M. *Electrophoresis* **1998**, *19*, 427.

[55] Righetti, P.G.; Conti, M.; Gelfi, C. *J. Chromatogr. A* **1997**, *767*, 255.

Separation of Enbrel® (rhuTNFR:Fc) Isoforms by Capillary Isoelectric Focusing

C. Jochheim / S. Novick / A. Balland / J. Mahan-Boyce / W.-C. Wang / A. Goetze / W. Gombotz

Department of Analytical Chemistry and Formulation, Immunex Corporation, 51 University Street, Seattle, WA 98101, USA

7.1 Abstract

Conditions that allow resolution of different isoforms of Enbrel® (rhuTNFR:Fc) by capillary isoelectric focusing (cIEF) and comparisons of electropherogram profiles with densitometer scans of gel IEF banding patterns are presented. The cIEF method uses internal standards and shows inter- and intra-assay reproducibility of relative migration with percent relative standard deviations < 2%. Isoform species whose isoelectric points (pIs) differ by 0.05 pI units are clearly resolved. The recently commercialized technology of imaged cIEF, a method performed without the mobilization step, is evaluated with TNFR:Fc. In comparison to the presented cIEF method, this method gives, as expected, a slightly different profile of the TNFR:Fc isoforms.

Several applications of the developed cIEF method illustrate its capacity to differentiate the complex charge heterogeneity of TNFR:Fc. Shifts in pI resulting from partial or complete desialylation with neuraminidase can be monitored by this CE method. Extended neuraminidase treatment collapses the complex isoform profile to two major and several minor forms. cIEF was used to analyze fractions from a preparative IEF separation, and to analyze samples after different process parameters were introduced. It is demonstrated here, that this method is well suited to monitor changes in product quality.

7.2 Introduction

Capillary isoelectric focusing (cIEF) can be used in place of slab gel IEF, with advantages including on tube detection and automated quantitation of the separated species [1]. As in gel IEF, proteins are separated according to their isoelectric point (pI) in a pH gradient formed by carrier ampholytes when an electric potential is applied. In CE the gel matrix is replaced by a viscous solution in the capillary. The most common method of detection is on-line UV detection after mobilization of the focused proteins past a detector window at a fixed point along the capillary [2]. In a different approach, Wu et al. [3] recently developed a new cIEF instru-

ment. With this method the mobilization step is omitted and a picture of the focused proteins along the whole capillary is taken with a CCD (charge coupled device) camera. A disadvantage of this approach is the lack of possible fraction collection and on-line mass-spectrometry detection of the different isoforms. However, there are two major advantages of this imaged cIEF technology. The first advantage is that there is no loss of resolution due to mobilization of the protein zones. Additionally the analysis is much faster, possibly resulting in a reduced development time. Applications of this technology have not been reported widely. We describe here initial results with TNFR:Fc isoforms, which provide a source of opti-

mism for the general utility of this technique.

CIEF is finding increasing use in analytical biotechnology. The uses of cIEF to determine isolectric points of proteins [4], to monitor thermal stability [5] and separation of protein drug from excipients in formulations [6] have been presented. Most commonly cIEF is now used for monitoring isoforms of biotechnology products [7–10]. Recently Hunt et al. [10] published a capillary isoelectric focusing method for resolving four species of a recombinant monoclonal antibody to human CD20 antigen. This assay is used as a test for identity and charge distribution, which results from the presence or absence of C-terminal lysine and possibly deamidation. This is the first cIEF assay that has been validated in accordance with the "International Conference on Harmonization" (ICH) guidelines [11]. Capillary isoelectric focusing also offers a fast assay to monitor the consistency of glycoforms produced in recombinant proteins. Variations in the glycosylation can have a significant effect on biological activity, clearance and stability [12]. The analysis of glycoproteins by cIEF typically produces a complex profile due to a heterogeneous mixture of isoforms with different oligosaccharides attached to the protein backbone.

We present here cIEF results obtained with recombinant human tumor necrosis factor receptor (p75) Fc fusion protein (rhu TNFR:Fc), Enbrel®, approved for the treatment of rheumatoid arthritis (RA) and juvenile rheumatoid arthritis (JRA). Tumor necrosis factor (TNF) is

0009-5893/00/02   59-07 $ 03.00/0

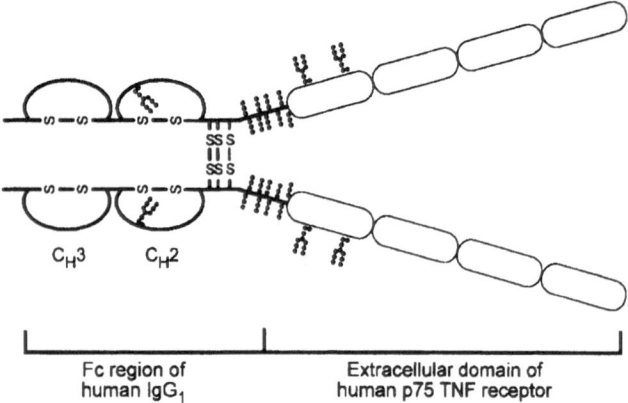

Figure 1. Schematic structure of recombinant human TNFR:Fc (Enbrel[R]). Enbrel[R] is a 467 amino acid containing homodimer, linked via three inter-chain disulfide bonds in the Fc (IgG) domain. Glycosylation sites and other disulfide linkages are indicated.

thought to be partly responsible for joint inflammation that leads to pain, erosion, and deformity as well as other systemic effects of RA and JRA. The Fc portion of the molecule provides a scaffold for dimerization of the tumor necrosis factor receptor. Dimeric TNFR is required for high affinity binding to the tumor necrosis factor. TNFR:Fc is a glycoprotein produced in Chinese hamster ovary (CHO) cells [13], and is composed of a chain of 467 amino acids. The homodimer contains three inter-chain disulfide bonds in the Fc domain (Figure 1). It includes 3 N-linked glycosylation sites and multiple O-linked glycosylation sites. All N-linked sites and multiple O-linked sites are occupied with typical mammalian oligosaccharide structures. These sugar structures are partially terminated with negatively charged sialic acid residues. Different amounts of sialic acid capping of oligosaccharides, attached to different sites on the protein backbone, lead to the very complex mixture of isoforms of TNFR:Fc. In this chapter, we demonstrate that cIEF is a useful tool for characterization and monitoring of TNFR:Fc isoforms.

7.3 Experimental

7.3.1 cIEF

Separations were carried out on a BioFocus 3000 instrument (Bio-Rad, Hercules, CA, USA). The capillary was a neutral coated capillary (eCAP) from Beckman, which is 24 cm in length, 50 μm I.D. × 375 μm O.D. with a distance of 19.4 cm from the inlet to the detector. The catholyte was 40 mM NaOH, the anolyte was 20 mM H_3PO_4 and the chemical mobilizer

was "Cathodic Mobilizer" from Bio-Rad Laboratories. For pI standards, Bio-Mark[TM] pI low molecular weight markers were used (Bio-Rad). If not mentioned otherwise, TNFR:Fc samples were diluted to 350 μg mL^{-1} in 2% ampholyte solution (Ampholines pH 3.5–pH 9.5, Pharmacia Biotech), 2.5% N, N, N', N'-tetramethylethylenediamine (TEMED, Pierce) and 0.2% methylcellulose. This mixture was then injected for 60 seconds at a pressure of 100 psi. Focusing time at 15 kV was 4 minutes and chemical mobilization took typically 31 minutes at 15 kV.

7.3.2 IEF Gel Analysis

Pre-cast IEF slab gels from Novex (pH 3–pH 7) were used unless noted otherwise. Electrophoresis was carried out for 2 h at 2 W (500 V limit). Gels were fixed using 12% trichloroacetic acid and stained with a colloidal Coomassie blue stain (Pro Blue, Owl Separations).

7.3.3 Desialylation

TNFR:Fc was treated with sialidase, isolated from *Arthrobacter Ureafaciens* (Oxford GlycoSciences). For complete desialylation 1 U sialidase / mg protein was incubated for 18 hours at 37 °C. Partial desialylation was achieved using 0.01 U enzyme / mg protein incubated for 10 minutes at ambient temperature.

7.3.4 Sialic Acid Determination

After enzymatic removal of sialic acid, TNFR:Fc samples were subjected to a

HPAEC-PAD (high performance anion exchange chromatography – pulsed amperometric detector) system from Dionex. A standard curve was prepared by adding 100 μg BSA to sialic acid in a final volume of 200 μL 2 M acetic acid and standards were treated the same as the samples. A 20 μL sample was separated on a Carbo-Pac PA-1 column (4 mm × 250 mm, Dionex) with a gradient of 100 mM NaOH (eluent A) and 100 mM NaOH containing 1 M NaOAc (eluent B). Over 20 minutes eluent B increased from 5% to 18%, was held at 18% for 1 minute, and then returned to initial conditions. The sialic acid content of the sample was determined by extrapolation on the standard curve.

7.3.5 Preparative IEF Fractionation

Isoforms of TNFR:Fc were separated using a Hoefer IsoPrime multi-chambered electrofocusing unit (Pharmacia). A specific pH for each membrane (separating the chambers) was generated by the use of a combination of acrylamido buffers (Fluka) of well defined pK values. Acrylamide solutions were prepared at each pH value by mixing the appropriate combination of acrylamido buffers with the acrylamide monomer and the cross linking reagent. Then the polyacrylamide gels were allowed to polymerize on glass microfiber filters (Whatman). TNFR:Fc membranes were prepared at the following pH values: 4.4, 4.55, 4.7, 4.85, 5.0, 5.15 and 5.3. TNFR:Fc was loaded in a single chamber (chamber 3, pI membranes pH 4.55 to pH 4.7) at a concentration of 0.5 mg mL^{-1}. Solubilizing agents were added in order to limit isoelectric point insolubility of the proteins and included 3 M urea/ 20% glycerol/2% CHAPS (TNFR:Fc). The total protein load was 15 mg. Electrophoresis limits were set at 3500 V, 20 mA, 4 W with a separation time of 48 hours at 4 °C. Before analysis, fractions were purified by Protein A affinity chromatography.

7.3.6 Imaged cIEF

The recently commercialized technology of imaged cIEF (Convergent Bioscience Ltd., Toronto, Ontario, Canada) is described by Mao and Pawliszyn [14]. Figure 2 shows the schematic of the imaged cIEF instrument iCE280. The separation

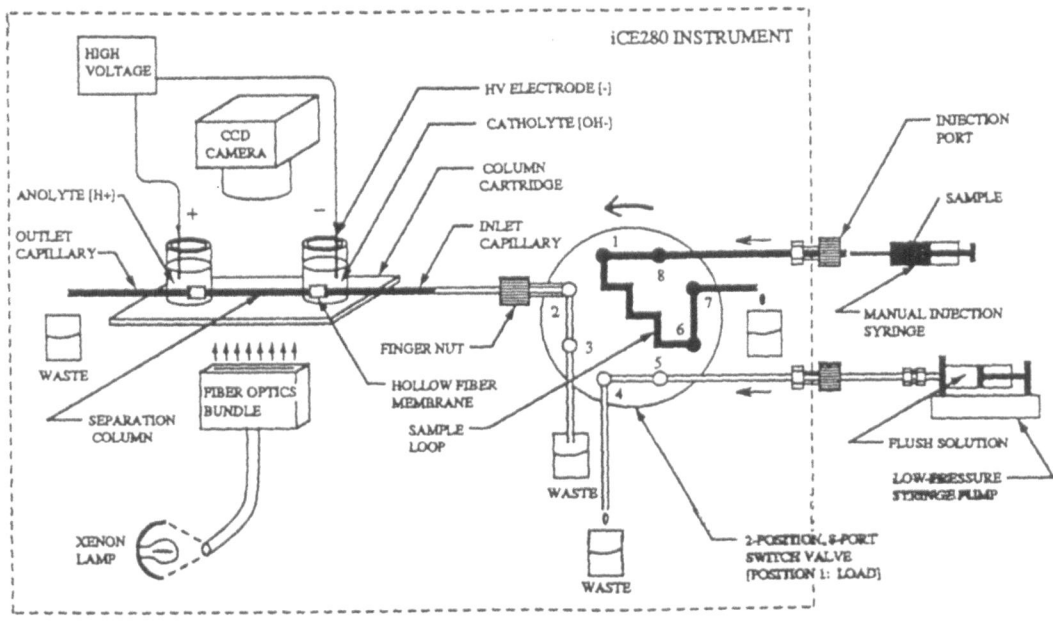

Figure 2. Block diagram of the imaged cIEF instrument, iCE280 (with permission from Convergent Bioscience, Ltd.). The switch valve is shown in load position. When switched to inject position, the syringe pump flushes the sample into the separation capillary, where focusing occurs. The CCD (charge coupled device) sensor captures the image of the separated species.

column is a $50\,mm \times 100\,\mu m$ I. D. $\times 200$ μm O. D. silica capillary with no outside coating (Polymicro Technologies, Tucson, AZ, USA). The inside of the column is coated with fluorocarbon to eliminate or reduce electroosmotic flow. The two ends of the column, as shown in Figure 2, are connected to two pieces of 3 mm long dialysis hollow fiber membrane (relative molecular mass cut-off: 18000). The two sections of the fiber are inserted into the electrolyte reservoirs. The anolyte solution was $100\,mM\ H_3PO_4$ and the catholyte solution was $40\,mM\ NaOH$. The column was conditioned with 0.35% methylcellulose. A TNFR:Fc sample was prepared at $500\,\mu g\,mL^{-1}$ with 4% carrier ampholyte (Servalyt pH 2–pH 11, Boehringer Ingelheim, Germany) in 0.35% methylcellulose. $15\,\mu L$ sample was injected and flushed into the separation column via the syringe pump, set at $5\,\mu L\,min^{-1}$. Focusing occurred at $3\,kV$ for 4 minutes. All separated isoforms were then recorded by the whole column absorption imaging detector (CCD camera) at 280 nm (Xenon Lamp).

7.4 Results and Discussion

7.4.1 cIEF Method to Separate TNFR: Fc Isoforms

Analysis of TNFR:Fc by slab gel IEF results in a profile composed of numerous bands spanning a pI range from about pI 4.2–pI 6.0. This profile appears far more complex than observed with most recombinant glycoproteins. A cIEF assay was developed which separates the isoforms of TNFR:Fc comparable to the gel. This method uses wide range ampholytes (Ampholines pH 3.5–pH 9.5), a neutral coated eCAP capillary and chemical mobilization. Three pI standards were added to normalize the migration times of the protein isoforms. Figure 3 shows a comparison of the densitometer scan of the Coomassie stained slab gel with the electropherogram of the separated isoforms of TNFR:Fc. The profiles are comparable with an enhanced resolution of some isoforms in the cIEF profile. Because no low pI marker was available to bracket the TNFR:Fc profile, the linearity of the pH gradient was tested with low molecular weight synthetic pI markers (Bio-Rad) spanning pI 4.6–pI 8.4. The linearity was acceptable with a correlation coefficient of 0.989. The pIs for the major TNFR:Fc peaks are determined by linear regression of the three standards. Typically a cathodic spacer, such as TEMED, is used if the sample contains basic components with pIs higher than pI 8 [15]. Although TNFR:Fc does not contain such basic species, TEMED was added resulting in better reproducibility of the migration times and sharper peaks (see Figure 4). Figure 5 shows electropherograms

of injections with different protein concentrations. Resolution is reduced significantly at $600\,\mu g\,mL^{-1}$, and precipitation was observed frequently at higher concentrations. TNFR:Fc injected at a concentration of $350\,\mu g\,mL^{-1}$ was found to exhibit an optimal signal to noise ratio, without loss in resolution. Table I shows the results of a reproducibility study of this assay. The major TNFR:Fc peak focuses around pI 4.8 with relative standard deviations (RSD) of less than 1.5%. Further separation of the complex isoform profile was achieved when narrow range ampholytes (Ampholines pH 3.5–pH 5.5) were blended with the wide range ampholytes (see Figure 6). However, since reproducibility of this approach was not as good, the cIEF method using the wide range ampholytes was applied to further experiments and characterization.

7.4.2 Imaged cIEF

Figure 2 shows the schematic of the newly developed approach of whole column detection, where the mobilization step is omitted. This method is described in detail by Mao and Pawliszyn [14]. Figure 7 shows the electropherogram of separated Enbrel isoforms as captured by the CCD camera at 280 nm after a 4 minute focusing step. Note: The x-axis indicates the distance between catholyte and anolyte, with low pI species at the left and high pI

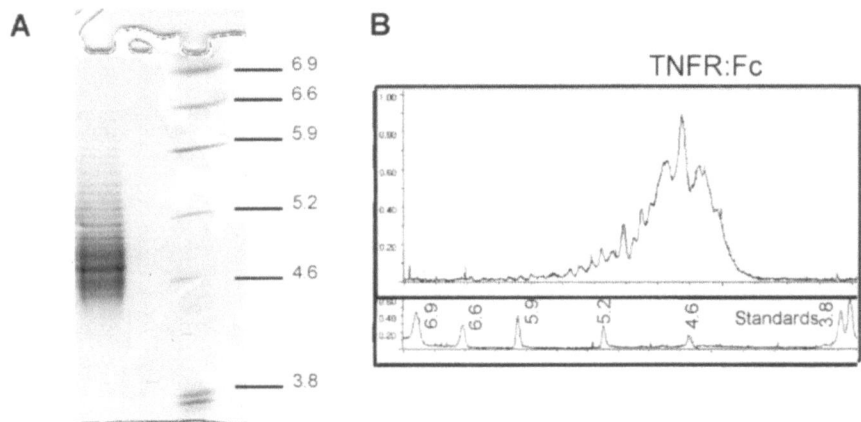

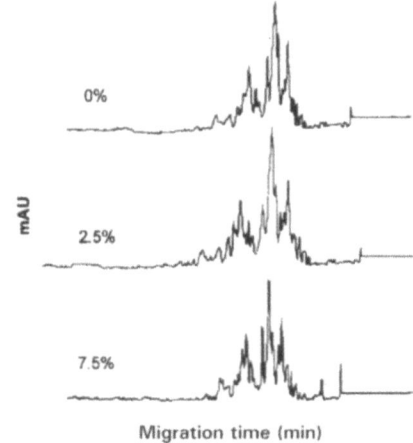

Figure 4. Effect of TEMED on the cIEF profile of TNFR:Fc. 0.35 mg mL^{-1} rhuTNFR:Fc was injected with 0%, 2.5% or 7.5% TEMED. Higher TEMED concentration helped reproducibility and resolution.

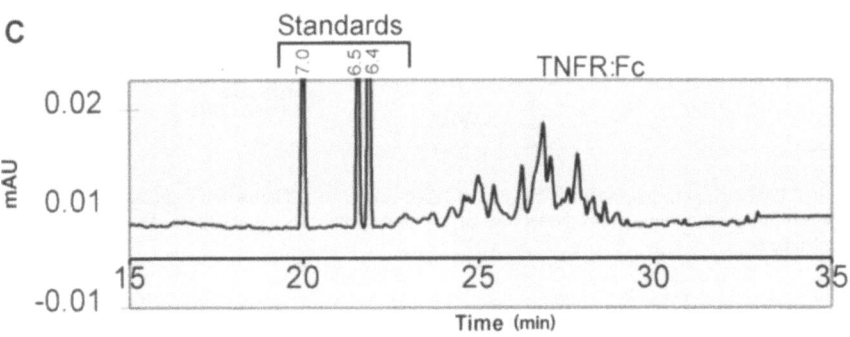

Figure 3. Comparison of the rhuTNFR:Fc IEF slab gel with the isoform profile generated by cIEF. Panel **A** shows a Coomassie stained IEF gel of rhuTNFR:Fc with standards (Novex pre-cast gel pI range 3.5–7.0). Panel **B** shows a densitometer scan of the IEF gel. Panel **C** shows a cIEF electropherogram of rhuTNFR:Fc injected at 0.35 mg mL^{-1} with 2% wide range ampholytes (Ampholine pH 3.5–9.5), 0.2% methylcellulose, and 2.5% TEMED.

species at the right side. This is the opposite direction than shown in conventional cIEF electropherograms. As seen in the cIEF described above, the imaged cIEF electropherogram shows three major and several minor isoform regions with more species at the high pI side of the profile. This technique appears very appealing to us, because of its short analysis time (6 minutes per sample). The development process can be reduced significantly and, since no mobilization step is required, this technique should potentially be more robust than cIEF assays run with conventional CE instrumentation. Another advantage we observed with this technique is less protein precipitation problems, again due to the short analysis time. However, one major disadvantage of this technology in its current format is the lack of on-line characterization of the separated species. Unlike conventional cIEF, imaged cIEF does not offer fraction collection, or interfacing with mass spectrometry. A combination of both methodologies should be most proficient.

7.4.3 Application of the Developed cIEF Method for Characterization of Enbrel® Isoforms

Treatment of TNFR:Fc with neuraminidase shifts the isoform profile towards basic in both the slab gel (Figure 8a) and cIEF (Figure 8b). While the isoelectric point of the protein backbone is 7.2, experimentally TNFR:Fc ranges in pI from pI 4.2–6.0. Partial release of sialic acid results in a gradual shift towards higher pI.

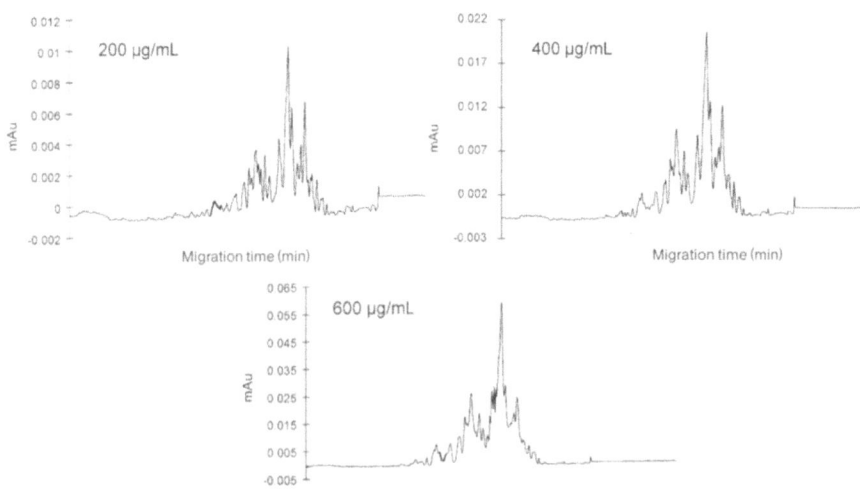

Figure 5. cIEF of TNFR:Fc at increasing concentrations from 200 μg mL^{-1} to 600 μg mL^{-1}. Optimal signal to noise ratio was found at 350 μg mL^{-1} without significant loss in resolution.

Table 1. Intra- and inter-assay reproducibility of C-IEF of TNFR:Fc. Assay 1: capillary had > 100 injections. Assay 2: capillary had > 300 injections. Assay 3: different capillary, unused.

Sample	Mean pI of major TNFR:Fc peak	% RSD
Assay 1: n = 20	4.80	0.51
Assay 2: n = 10	4.86	0.92
Assay 3: n = 10	4.73	0.12
Inter-assay: n = 3	4.80	1.12

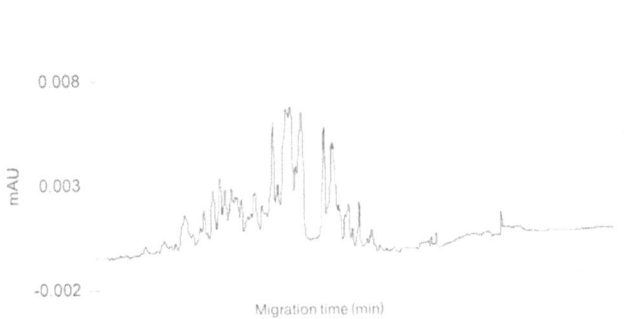

Figure 6. cIEF electropherogram of TNFR:Fc injected at 350 μg mL⁻¹ with 2% preblended narrow range ampholytes (Ampholine pH 3.5–5.5) with wide range ampholytes (Ampholine pH 3.5–9.5), 0.2% methylcellulose, and 2.5% TEMED.

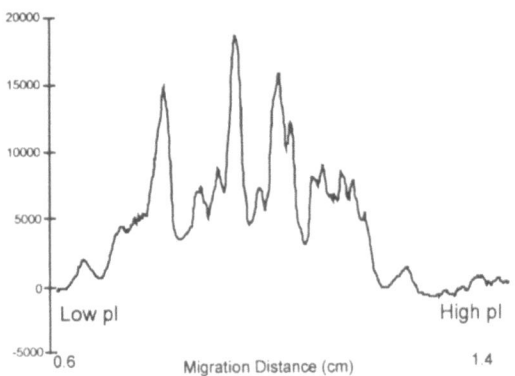

Figure 7. cIEF of TNFR:Fc with whole column detection. Recombinant human TNFR:Fc was injected at 500 μg mL⁻¹ with 4% ampholytes (Servalyte pH 2–11) and 0.35% methylcellulose. Focusing occurred at 3 kV for 4 minutes.

After extended desialylation, only two major and four minor species are focused around pI 7.2. This is clearly seen in the gel, and in the electropherogram. Analysis of the neuraminidase treated samples for residual sialic acid indicated a loss of four sialic acids after a mild sialidase treatment, and complete desialylation after excessive enzyme treatment (Figure 8b). This indicates, that the complexity of the TNFR:Fc preparation is mostly due to heterogeneity in sialic acids attached to the oligosaccharide residues. Identification of the remaining species after complete removal of sialic acid remains to be elucidated. cIEF could also monitor the consistency of the different desialylated TNFR:Fc isoforms, but this remains to be investigated.

Preparative IEF in a liquid vein using immobilized pH gradients [16] has been successfully used to characterize isoforms of several recombinant proteins [17, 18]. This technique was used here to prepare sizable amounts of TNFR:Fc isoforms. The separation of isoforms showed some variation depending on the choice of solvents. Glycerol, urea and CHAPS were used to help maintain the isoforms in solution at their isoelectric point and improve protein transport across membranes. The best result was obtained with a concentration of 3 M urea / 20% glycerol / 2% CHAPS. After collection, the fractions were further purified by protein A affinity chromatography. Analysis of the collected fractions on analytical IEF gel and cIEF showed that the TNFR:Fc isoforms have been separated into several defined regions (Figure 9). Due to the high number of isoforms, visualized by analytical IEF in the original molecule, a complete separation of each individual species was unlikely,

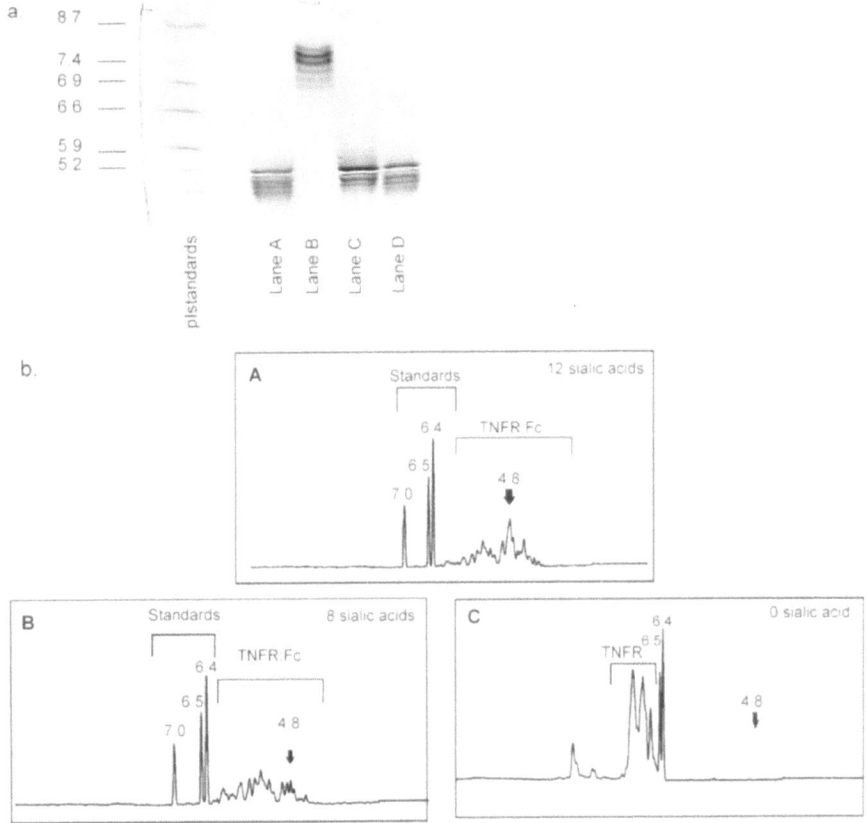

Figure 8. Desialylation of rhuTNFR:Fc. **a** Slab gel IEF (Novex pre-cast, pH 3–10). **b** cIEF under conditions as described in the text, with Ampholines, pH 3.5–9.5. Lane A/ Panel A: untreated rhuTNFR:Fc held at 4 °C. Lane B/ Panel C: rhuTNFR:Fc treated with 1 U sialidase /mg protein for 18 hours at 37 °C. Lane C/Panel B: rhuTNFR:Fc treated with 0.01 U sialidase /mg protein for 10 min at ambient temperature. Lane D: Untreated rhuTNFR:Fc held at 37 °C for 18 hours. Results from a sialic acid determination are indicated in **b**.

and could not be achieved. The cIEF electropherograms of these fractions (Figure 9b) appear to show more species than observed in the gel (Figure 9a), indicating a slightly better resolution in the CE separation. After purification by protein A affinity chromatography, the separated fractions were analyzed for biological activity and in a receptor-ligand binding assay.

Bioassay and binding assay indicated that the different regions had similar bioactivities (Data not shown). Sialic acid compositional analysis showed that the microheterogeneity was linked to a variation of sialic acid content of about 4 moles sialic acid/ mole TNFR:Fc between the acidic and basic regions of the cIEF profile (indicated in Figure 9b). Preparative separa-

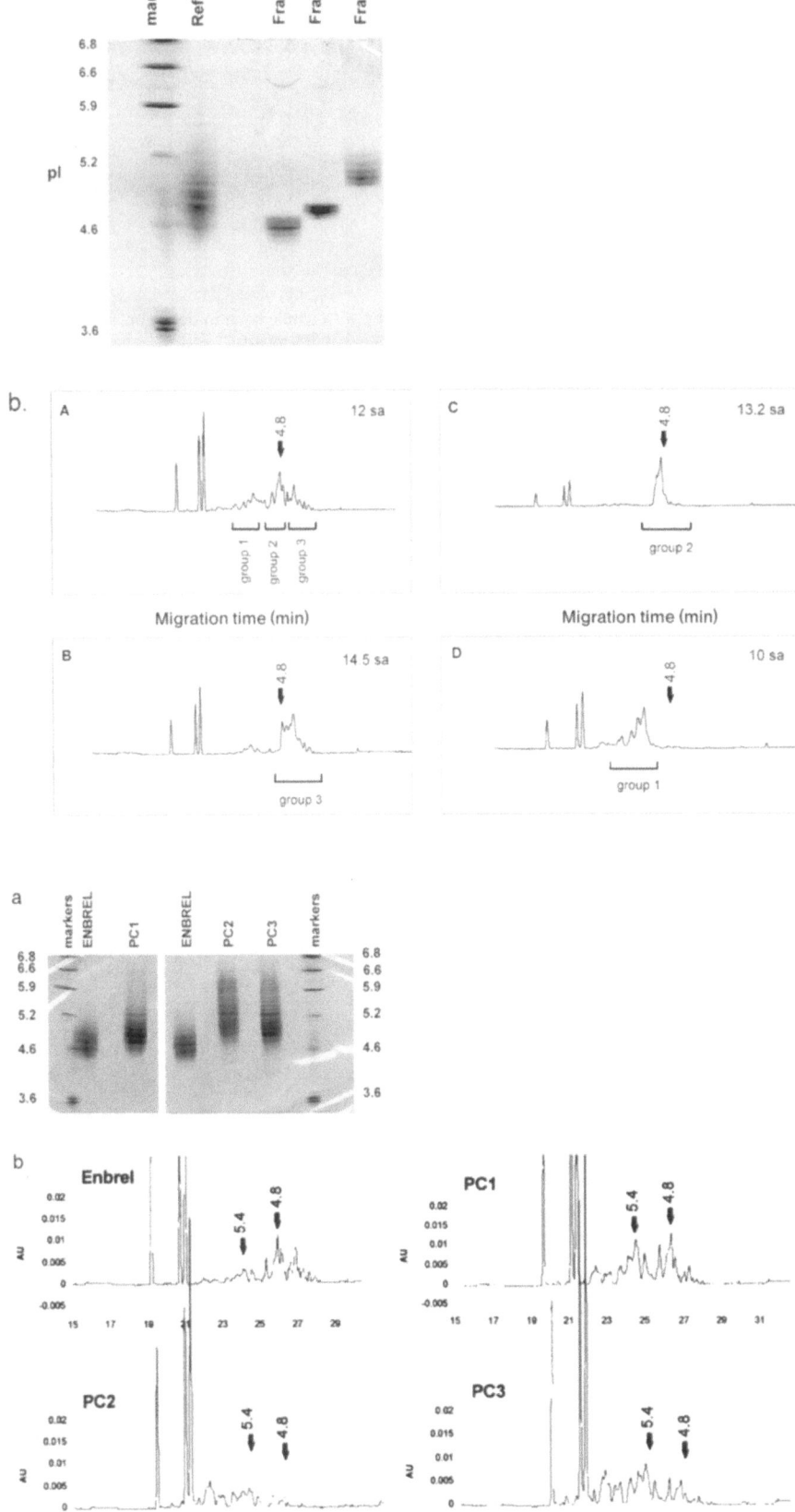

◁ **Figure 9.** Analysis of rhuTNFR:Fc fractions from a preparative scale separation with a Hoefer IsoPrime unit. Isoforms of rhuTNFR: Fc were partially separated on an immobilized pH gradient. Three fractions (B, C, D) were purified and analyzed on a Novex pre-cast slab gel IEF **a** and by cIEF **b** and compared to Enbrel (A). Results from a sialic acid (sa) determination are indicated in **b**.

tion of TNFR:Fc isoforms proves that the distribution of microheterogenities, visualized by cIEF, has no significant difference in the final potency of the product.

Lastly, the cIEF method was tested for its ability to monitor changes in the isoform distribution, when different process parameters were introduced in the TNFR:Fc manufacturing process. The slab gel IEF (Figure 10a) as well as the cIEF electropherograms (Figure 10b) show a shift of the isoform profile to higher pI, when compared with TNFR:Fc. Careful analysis of the cIEF results indicate, that when process changes were made, the isoform identities remained the same, but the relative distribution shifted in favor of isoforms with higher pI. In contrast to standard slab gel IEF, this cIEF method with the resolution of 0.05 pI units provides a sensitive method for detection of potential new species generated as a result of process changes.

We have shown, that the separation of TNFR:Fc isoforms by cIEF is comparable to the separation by IEF slab gels. Although complete separation of all isoforms of this complex glycoprotein profile could not be achieved, the cIEF method is well suited to monitor changes in isoform distribution due to change in sialic acid content. In conclusion, the cIEF methods presented in this study for the separation of TNFR:Fc isoforms provide a valuable asset for evaluating product quality and process consistency and are now in routine use in process development.

7.5 Acknowledgements

We would like to thank Dr. Jiaqi Wu from Convergent Bioscience Ltd. for the demonstration of the imaged cIEF technique, and the generation of the TNFR:Fc data on this instrument.

Figure 10. Monitoring process changes of rhuTNFR:Fc. A comparison of slab gel IEF (**a**) and cIEF (**b**). Changes were introduced in the production of rhuTNFR:Fc (PC1, PC2, PC3) and product quality was monitored by IEF.

7.6 References

[1] Hjerten, S.; Elenbring, K.; Kilar, F.; Liao, J.L.; Chen, A.J.; Siebert, C.J.; Zhu, M.D. "Carrier-free zone electrophoresis, displacement electrophoresis and isoelectric focusing in a high-performance electrophoresis apparatus." *J. Chromatogr.* **1987**, *403*, 47–61.

[2] Pritchett, T.J. "Capillary isoelectric focusing of proteins." *Electrophoresis* **1996**, *17*, 1195–1201.

[3] Wu, J.; Li, S.C.; Watson A. "Optimizing separation conditions for proteins and peptides using imaged capillary isoelectric focusing" *J. Chromatogr. A* **1998**, *817*, 163–171.

[4] Liu, X.; Sosic, Z.; Krull, I.S. "Capillary isoelectric focusing as a tool in the examination of antibodies, peptides, and proteins of pharmaceutical interest." *J. Chromatogr. A* **1996**, *735*, 165–190.

[5] Dai, H.J.; Krull, I.S. "Thermal stability of immunoglobulins using capillary isoelectric focusing and capillary zone electrophoretic methods." *J. Chromatogr. A* **1998**, *807*, 121–128.

[6] Park, S.S.; Sloey, C.J.; Chang, B.S. "Applications of Capillary Electrophoresis (CE) to Formulation Development of Amgen Proteins." presented at CE in Biotechnology: Practical Applications for Protein Analysis, San Francisco (August 18, **1999**).

[7] Moorhouse, K.G.; Eusebio, C.A.; Hunt, G.; Chen, A.B. "Rapid one-step capillary isoelectric focusing method to monitor charged glycoforms of recombinant human tissue-type plasminogen activator." *J. Chromatogr. A* **1995**, *717*, 61–69.

[8] Hagmann, M.L.; Kionka, C.; Schreiner, M.; Schwer, C. "Characterization of the F(ab')2 fragment of a murine monoclonal antibody using capillary isoelectric focusing and electrospray ionization mass spectrometry." *J. Chromatogr. A* **1998**, *816*, 49–58.

[9] Lee, H.G. "Rapid high-performance isoelectric focusing of monoclonal antibodies in uncoated fused-silica capillaries." *J. Chromatogr. A* **1997**, *790*, 215–223.

[10] Hunt, G.; Hotaling, T.; Chen, A.B. "Validation of a capillary isoelectric focusing method for the recombinant monoclonal antibody C2B8." *J. Chromatogr. A* **1998**, *800*, 355–367.

[11] International Conference on Harmonization. Guideline on validation of analytical procedures: definition and terminology. Fed. Reg. 60 (40) (March 1, **1995**) 11260–11262.

[12] Boyd, P.N.; Lines, A.C.; Patel, A.K. "The effect of the removal of sialic acid, galactose and total carbohydrate on the functional activity of campath-1H." *Mol. Immunol.* **1995**, *32*, 1311–1318.

[13] Mohler, K.M.; Torrance, D.S.; Smith, C.A.; Goodwin, R.G.; Stremler, K.E.; Fung, V.P.; Madani, H.; Widmer, M.B. "Soluble tumor necrosis factor (TNF) receptors are effective therapeutic agents in lethal endotoxemia and function simultaneously as both TNF carriers and TNF antagonists." *J. Immunology* **1993**, *151*, 1548–1561.

[14] Mao, Q.L.; Pawliszyn, J. "Capillary isoelectric focusing with whole column detection for analysis of proteins and peptides." *J. Biochem. Biophys. Methods* **1999**, *39*, 93–110.

[15] Wehr, T.; Rodriguez-Diaz, R.; Zhu, M. "Capillary Electrophoresis of proteins." *Chromatographic Sciences Series* **1999**, *80*, 131–233.

[16] Righetti, P.G.; Bossi, A. "Isoelectric focusing in immobilized pH gradients – recent analytical and preparative developments." *Anal. Biochem.* **1997**, *247*, 1–10.

[17] Righetti, P.G.; Bossi, A.; Wenisch, E.; Orsini, G. "Protein purification in multicompartment electrolyzers with isoelectric membranes." *J. Chromatogr. B* **1997**, *699*, 105–115.

[18] Balland, A.; Mahan-Boyce, J.A.; Krasts, D.A.; Daniels, M.; Wang, W.; Gombotz, W.R. "Characterization of the isoforms of PIXY321, a granulocyte-macrophage-colony stimulating factor – interleukin-3 fusion protein, separated by preparative isoelectric focusing on immobilized pH gradients." *J. Chromatog A* **1999**, *846*, 143–156.

Optimization, Validation, and Use of Capillary Gel Electrophoresis for Quality Control Testing of Synagis®, a Monoclonal Antibody

M. A. Schenerman / S. H. Bowen

MedImmune, Inc., 35 W. Watkins Mill Rd., Gaithersburg, MD 20878, USA

8.1 Abstract

Capillary electrophoresis methods offer many advantages over their conventional electrophoresis counterparts including: greater precision, higher throughput, on-line detection and less waste. Capillary gel electrophoresis (CGE), using replaceable sieving polymers, is widely used as an alternative to SDS-PAGE and densitometry. CGE offers greater automation capabilities than SDS-PAGE and can be operated using conventional chromatography system software. This software for handling and analyzing data makes the validation of CGE methods look quite similar to the validation of an HPLC method. The same parameters can be examined, including: precision, accuracy, linearity, selectivity, range, robustness, and system suitability. There are, however, certain key elements that are critical to a successful CGE method validation. The CGE quantification method must be shown to be equivalent to SDS-PAGE with densitometry and should be presented such that the densitometry plots look like the electropherograms. In addition, the sensitivity of the methods must be compared and the CGE method must have comparable or better sensitivity to SDS-PAGE with Coomassie staining. The validation must also be performed using materials that might be tested during the course of a typical process (in-process samples) in their particular matrices. The other vital element is assuring that acceptance criteria for the validation have been set to reasonable ranges. These ranges must be determined in pilot studies that are executed prior to the validation studies. Under most CGE conditions, the acceptance criteria will be quite similar to those seen in HPLC validations. Overall, CGE methods can be validated to the same level of assurance as HPLC methods. The CGE method described here has been approved by the FDA and is used routinely for testing of Synagis®.

8.2 Introduction

Capillary electrophoresis (CE) is a separation technique that can be applied to biopharmaceuticals using a wide variety of formats. For example, proteins can be resolved using Capillary Zone Electrophoresis, Capillary Isoelectric Focusing, Affinity Capillary Electrophoresis, Capillary Electrochromatography, and Capillary Gel Electrophoresis (CGE) which is sometimes referred to as "non-gel sieving" (CE-NGS). All of these separation formats have been reviewed previously [1–5].

Regardless of the separation mechanism, CE methods offer many advantages over their conventional HPLC or gel counterparts [6–10]. CE methods typically use narrow bore (25–75 μm inner diameter) fused silica capillaries with high voltages (10–30 kV) applied during the separation. The high resistance of the capillary limits Joule heating and results in a high efficiency ($N > 10^5$ to 10^6) separation

and short analysis time. Typically, small sample volumes (1–50 nL) are applied to the capillary. Commercial CE systems have automatic temperature control and can introduce sample injections using an autosampler. Usually CE systems use aqueous buffers in relatively small volumes meaning there is a limited amount of waste (particularly organic wastes like methanol and acetic acid used for slab gel staining and destaining that are expensive to dispose of and potentially hazardous).

Capillary gel electrophoresis (CGE) is a means of separating proteins by size using the same sieving mechanism as SDS-PAGE [11–29]. Proteins are treated with SDS in the presence or absence of a reducing agent resulting in a complex with uniform mass to charge ratio [30]. The protein-detergent molecules are then separated in a high voltage electric field in the presence of a replaceable polymer sieving matrix. The polymer matrix entangles larger protein-detergent complexes and slows their migration through the capillary. In this way, smaller proteins migrate past the detector window first followed by proteins of progressively larger mass and hydrodynamic radius [27, 29].

Numerous investigators have attempted to use different polymeric sieving matrices for CGE including cross-linked polyacrylamide [11, 12], non-cross-linked linear polyacrylamide [13–15, 17], dextran [14, 19], pullulan [20], polyethylene oxide [22], and poly(vinyl) alcohol [25]. These authors have also used several techniques to determine molecular weights of proteins including plotting migration time against molecular weight standards [11–

0009-5893/00/02   66-09 $ 03.00/0

30] and the use of Ferguson plots [12, 16, 21, 22, 24].

This paper will illustrate the development, optimization, and validation of a CGE method used to support commercial production of Synagis® (a monoclonal antibody to prevent Respiratory Syncytial Virus infection in high-risk infants). The paper will point out the benefits and potential pitfalls of using a CE method in Quality Control testing. The paper will also suggest ways to simplify validation and streamline the required studies to receive regulatory approval.

8.2.1 Advantages of CGE over SDS-PAGE

CGE has many advantages over conventional SDS-PAGE using Coomassie staining and quantitation by densitometry. Some of these advantages are summarized on Table I. CGE offers substantial advantages in automation, data processing, waste generation, and labor savings. For example, because the data systems are frequently identical to HPLC software, data processing is faster and more quantitative. Conventional SDS-PAGE gels generate significant volumes of organic waste (Coomassie dye, methanol, and acetic acid) during the staining and destaining process. These organic waste materials are cumbersome to store, costly to dispose of, and potentially hazardous.

8.3 Methods

8.3.1 SDS Polyacrylamide Gel Electrophoresis

Recombinant proteins are diluted into sample dilution buffer containing SDS in the presence or absence of 5% β-mercaptoethanol (BME). The resulting samples are heated at 80 °C, cooled, and electrophoresed on a 4–15% polyacrylamide gradient gel. The gels are then Coomassie Blue R-250 stained, destained, and densitometry performed to compare with the electrophoretic pattern of the Reference Standard [30].

8.3.2 Capillary Gel Electrophoresis (CGE)

Recombinant proteins are diluted into sample dilution buffer (Bio-Rad catalog

#148–5033) in the presence or absence of 5% BME. The resulting samples are heated in a boiling water bath for 10 minutes and cooled. The capillary is then filled with CE-SDS Protein Run buffer (Bio-Rad Catalog #148–5032) and the sample is introduced using electrokinetic injection at – 10.0 kV for 40 seconds. The separation takes place in an electric field of – 390 V cm⁻¹ for 22 minutes at 50 °C in a Hewlett-Packard extended light path fused silica capillary {50 μm inner diameter (ID), 38.5 cm total length, 30 cm effective length} with detection at 220 nm on a Hewlett-Packard HP3DCE instrument [29]. Benzoic acid is used an internal standard in every run.

8.3.3 MALDI-TOF Mass Spectrometry

Aliquots (50–100 μg) of Synagis® are applied to Bio-spin 6 (Bio-Rad Laboratories) size-exclusion columns equilibrated in 2% TFA according to the manufacturer's protocol. Desalted samples (1 μL) are spotted in duplicate onto 20-well sample slides followed by an overlay of 0.5 μL of matrix (saturated sinapinic acid in 70% acetonitrile/30% of 0.1% TFA). Mass data (m/z) are collected on a Kratos Kompact MALDI 1 MALDI-TOF mass spectrometer using BSA as the external mass calibrant. Sample wells are scanned at 50 shots/well with laser power varying through the middle third of the range. Regions of high ion intensity are re-

scanned and the spectra stored. Each stored spectrum is the average of 50–100 shots.

8.4 Regulatory Guidance

There is considerable published guidance on analytical method validation [31–43]. The USP has now also published the harmonized ICH guidance documents in USP 24. In addition, there are similar guidance documents published by each of the pharmacopoeias worldwide (e. g, European Pharmacopoeia). While the United States, the European Union, and Japan have harmonized their approach to method validation, other countries may have their own unique requirements that must be fulfilled in order to achieve registration. A growing number of pharmaceutical quality control laboratories have been reporting the validation and routine use of CE assays [44–54].

Most validation studies will consist of an examination of the following parameters: specificity, linearity, range, accuracy, precision, limit of detection, limit of quantitation, robustness, and system suitability. The parameters selected for the validation study are dependent on the purpose of the analytical method. For example, for Limit Assays, only specificity and limit of detection parameters are required to be examined [35].

Specificity is the ability of the method to distinguish the molecule being analyzed from other similar contaminants that

Table I. Advantages and Disadvantages of CGE Over Traditional SDS-PAGE

Parameter	CGE	SDS-PAGE
System preparation	30 minutes	30 minutes
Sample preparation and loading	1 hour	30 minutes
Maximum number of samples	36 samples per tray	13 samples per gel
Separation	30 minutes per sample	1.75 hours per gel
Detection	Direct-based on UV detection or LIF following dye binding or derivatization	Indirect – 24 hours after separation (1 hour of fixing the gel and overnight Coomassie staining. This is followed by 6 hours of destaining gel priors to densitometry).
Digitization and data analysis	Included during the separation	1 – 2 hours
Waste generation	Minimal (milliliters)	Substantial (liters) containing solvents and dyes
Analysis throughput	108 samples per 3 day period	13 samples per 3 day period
Operator time	Automated	Labor intensive

Adapted from Reference [6].

might be found in the preparation. Linearity establishes a defined mathematical relationship between the instrument response and the analyte concentration (it need not always be linear). Range defines the concentrations of analyte between which the method can be used for quantification. Accuracy demonstrates how close the results from the test method are to the "true" values.

Precision is defined by three subcategories that together describe the "closeness" of values to each other. The first subcategory is repeatability, which describes the variability associated with the method when it is repeated under exactly identical conditions. The second subcategory is intermediate precision, which covers broader aspects of variability associated with instrument to instrument, day to day, and run to run precision. Intermediate precision is considered by many to be the true variability associated with the test method. The final subcategory is reproducibility, which defines the variability between laboratories.

Limit of detection and limit of quantitation define the limits at which analyte can be seen and reproducibly measured, respectively. Robustness describes a number of variables that may affect the performance of the method including buffer pH, ionic strength, voltage, etc. Finally, system suitability describes tests used to demonstrate that the method is operating properly at the time of the analysis. Test results are not considered acceptable without having met the system suitability criteria.

8.5 Optimization Strategy

The optimization of an analytical method should be a systematic process with an eye toward the ultimate validation studies. Often the optimization phase is given insufficient time and the method is not fully understood before the validation studies must be initiated. This will ultimately lead to disaster. A well-organized systematic optimization study will have two goals. The first goal is to "qualify" the method by showing reasonable linearity and precision. Frequently, spike-recovery studies are an integral part of method qualification. These studies demonstrate that the actual sample matrix, which is being analyzed, does not interfere with the quantification of the analyte – a common problem for biologicals. The second goal is to understand the factors that can contribute to variability in the assay. Certain questions should also be addressed during the optimization studies such as:

- What is the purpose of the test?
- Does the assay replace an existing gel system?
- What sensitivity must you achieve?
- What type of samples do you intend to analyze?
- Do you anticipate using this test for regulatory submissions?

The answers to these questions will show the path that should be followed toward efficient optimization studies. Many analytical methods do not need to be validated as they only support preclinical or early clinical studies. Whether or not the method replaces an existing method will also define how the studies should be done. The new method must be at least as good as, if not better than, the existing method. Knowing the required sensitivity is essential in determining the detection system used. For example, UV detection with CE is probably only as sensitive as Coomassie Blue-staining in gels. The types of samples to be analyzed will also have a major impact on the optimization studies. If the samples are crude in-process fractions (e. g, cell lysate), the complexities of the method increase dramatically. If the test is to be included in a regulatory filing (e. g, IND), the studies described above would be considered a minimum.

Optimization studies should be documented in a Development Report or a Method Optimization Report. Regardless of the format, the purpose of the report is to define the scientific rationale behind the experiments used to optimize the method. Each experiment performed should have a description of the purpose, materials used, method description, results, and a conclusion. Frequently, when the key person responsible for an optimization study leaves the company, the experience gained goes with that person. The optimization report is a valuable tool which can be utilized later in the product's evolution to retain the knowledge gathered during these studies.

8.6 Validation

8.6.1 Pilot Studies

Once the optimization studies have been completed and a reasonable understanding of the method has been gained, validation pilot studies should be initiated. These pilot studies should be designed and performed to determine validation criteria and critical parameters. The experiments should help to gain some further understanding of Intermediate Precision, Reproducibility, and Robustness. In addition, these studies can provide some insight into the stability of critical reagents. Pilot studies should also be used to reexamine the performance of the method with different in-process samples. Ultimately, these studies will determine which of the validation parameters are critical to the reliability of the method and will dictate what the acceptance criteria should be for each parameter.

8.6.2 Protocol/Execution

A validation protocol must be a pre-approved (by the Quality Unit) prospective document with defined acceptance criteria for each parameter. It must be sufficiently detailed to define the experiments unambiguously. The protocol should define the purpose, scope, and provide sufficient background so that even someone not familiar with the method could understand the rationale. The sections of the protocol should be organized based on the relevant ICH validation parameters. Each section must have clearly designated and supportable acceptance criteria.

The protocol should define how data will be collected and stored (e. g, validation binders, field notebooks, etc.). In addition, it should describe what to do in the event of an out-of-specification (OOS) result or a protocol deviation.

8.7 Validation Parameters for CGE

Validation of CGE methods is not much different from the validation of HPLC methods. There have been several published reports of validation studies for CGE [45, 47, 49, 51, 53]. Listed below are some example approaches to CGE validation, many of which parallel HPLC methods.

8.7.1 Specificity

Specificity can be demonstrated by comparing different proteins that may be

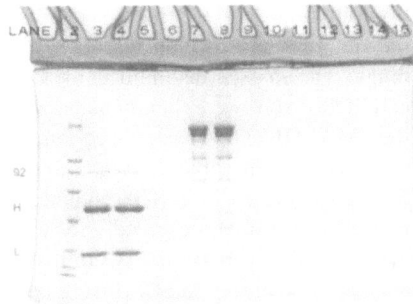

Figure 1. Reducing and non-reducing SDS-PAGE gel of Synagis® (Coomassie Blue-stained). Synagis® was diluted into sample dilution buffer containing 2% SDS with (reducing) or without (non-reducing) 5% (-mercaptoethanol, heated for 10 minutes, then loaded onto a 4–15% SDS-PAGE gel (3 μg protein/lane). Lane assignments were: Bio-Rad high and low molecular weight markers (lane 2), reduced Reference Standard (lane 3), drug substance (lane 4), and non-reduced recombinant Mabs: Reference Standard (lane 7), drug substance (lane 8). Following electrophoresis at constant voltage, the gel was fixed and Coomassie Blue-stained. Molecular weight markers are: myosin (200,000 Da), (β-galactosidase (116,250 Da), phosphorylase b (97,400 Da), bovine serum albumin (66,000 Da), ovalbumin (45,000 Da), carbonic anhydrase (31,000 Da), and soybean trypsin inhibitor (21,500 Da) (Adapted from [6]).

found in the manufacturing process. For example, bovine serum albumin (BSA) is a component in the cell culture process used to make Synagis®. It would, therefore, be reasonable to show that the CGE method could discriminate Synagis® from BSA. In addition, spike-recovery studies should be done on process samples from various points in the process, depending on how the method is used (e. g, after each purification step). The acceptance criteria should state that the BSA must not change product peak migration time but must change the product purity (decreased purity because of increased contaminant).

The same approach could be taken with insulin, another process residual. Spike-recovery studies can show the method is capable of detecting the contaminant if it were there (see Figure 3). The fact that these studies can be done using CGE shows an immediate benefit over SDS-PAGE because insulin would wash out of conventional SDS-PAGE gels.

Another important comparison to be made is to demonstrate that CGE is equally capable of detecting and quantifying proteins as SDS-PAGE with densitometry. A typical Coomassie Blue-stained SDS-PAGE gel of Synagis® is shown in Figure 1. Under reducing conditions, the

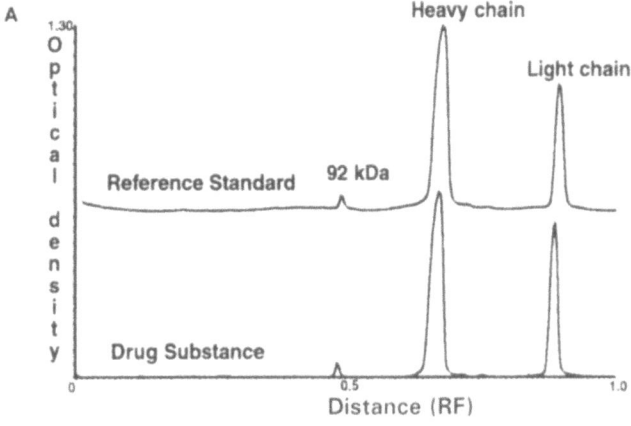

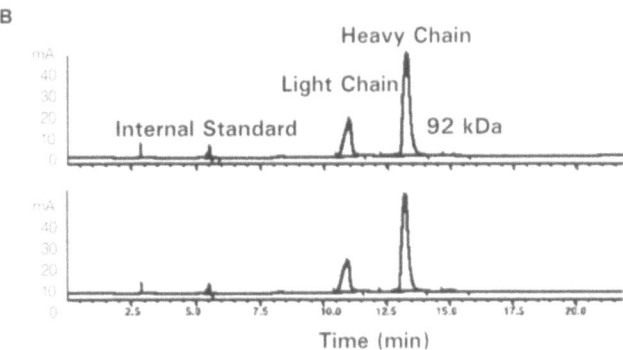

Figures 2. Comparison of Synagis® separated by reducing SDS-PAGE and reducing CGE. (**A**): SDS-PAGE Densitometry. Synagis® Reference Standard and Bulk Drug Substance were diluted into sample dilution buffer containing SDS in the presence of β-mercaptoethanol. The resulting samples were analyzed by SDS-PAGE as described in the Methods section. Following staining and destaining, the gel was digitized and densitometrically scanned using a cooled charge-coupled device camera and MCID software. (**B**): CGE. The same samples were analyzed by reducing CGE as described in the Methods section (Adapted from [6]).

92 kDa band, heavy chain, and light chain are clearly visible. When this same gel is digitally imaged and the band patterns quantified using densitometry the results are shown in Figure 2A. A similar pattern is seen when the same samples are analyzed by CGE (Figure 2B) but the order of peaks is reversed because the lowest molecular weight bands will move fastest through the capillary in CGE whereas the SDS-PAGE gel is scanned from top to bottom (hence the highest molecular weight bands are seen first). This type of direct comparison is vital in assisting the regulatory authorities to understand the value of the CGE assay.

8.7.2 Linearity

If the method is only used for identity testing or molecular weight determination, then the linearity parameter is probably not required. If, however, the method is used for purity assessment, then some consideration should be given to linearity or response.

Establish a linear correlation with at least six data points within the operating range. Use reasonable acceptance criteria based on 95% confidence limits (e. g, greater than or equal to 0.995 correlation). Perform enough replicates at each concentration to ensure statistical significance (e. g, triplicates).

8.7.3 Range

As mentioned above for the linearity parameter, if the method is only being used for identity testing, it is probably not necessary to validate linearity. For quantitative applications, the relevant range (50–150% of the target) is all that is required. The relevant range should include any in-process fractions that are anticipated being tested using the CGE technique. Don't overvalidate – it is not required to show the entire operating range of the detector.

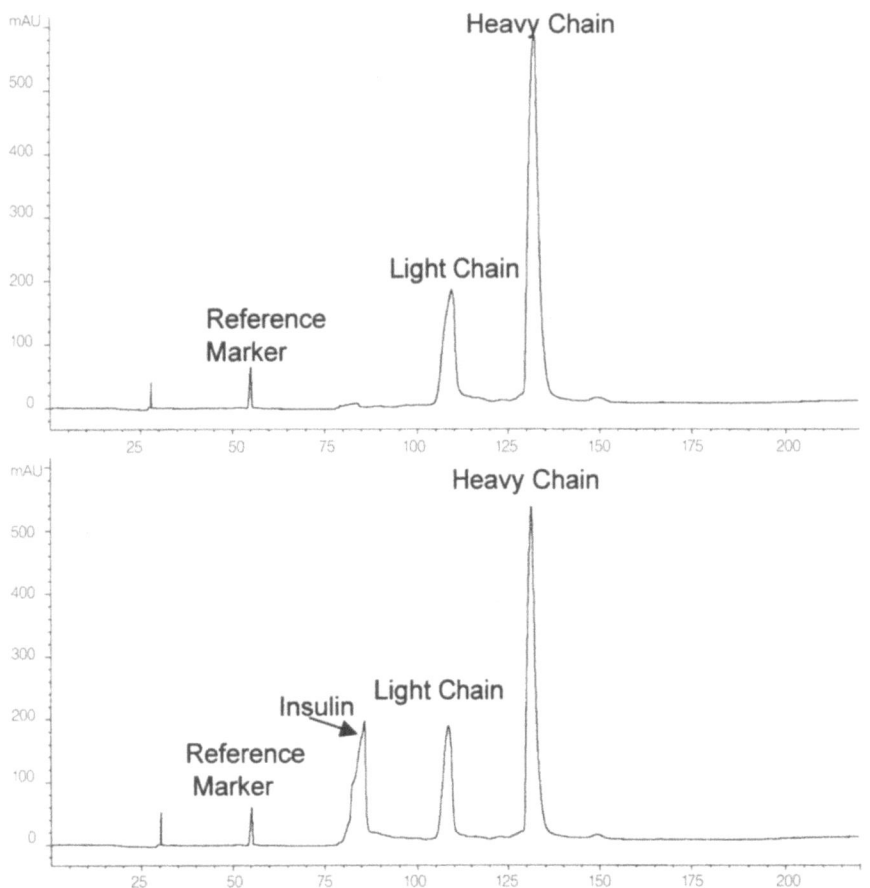

Figure 3. Specificity of CGE to discriminate between different polypeptide sizes. Hewlett-Packard extended light path capillary, 50 µm × 38.5 cm (30 cm to detector); sample buffer, CE-SDS Sample Buffer (Bio-Rad Laboratories) in the presence of β-mercaptoethanol buffer, CE-SDS Protein Run Buffer (Bio-Rad Laboratories); load, −10.0 kV for 40 sec (2 mg mL⁻¹ protein concentration); run, −15 kV; detection signal 220 nm. Graphic label assignments for CGE were: reduced Synagis® (top graphic), and reduced Synagis® spiked with 1 mg mL⁻¹ insulin (bottom graphic).

8.7.4 Accuracy

Frequently the accuracy of a method is evaluated by comparing the results to a "reference method". For SDS-PAGE or CGE a reference method to determine the mass of protein bands is matrix-assisted laser desorption time-of-flight (MALDI-TOF) mass spectrometry. Compare molecular weight determination of the heavy chain and the light chain using SDS-PAGE with densitometry, CGE, and

MALDI-TOF mass spectrometry in pilot studies to understand what the relative differences are. Set limits in the validation protocol based on the knowledge gained from those pilot studies.

Table II shows a validation study where the acceptance criteria was set at less than or equal to 20% difference between the molecular weight determinations for each of the methods compared to MALDI-TOF mass spectrometry. The results show that the % difference in mass

between MALDI-TOF mass spectrometry and CGE was within the acceptable range. The masses determined by CGE and SDS-PAGE were slightly higher than the MALDI-TOF mass spectrometry mass. These slight differences in mass may be due to differences in SDS binding per unit mass of protein, especially for alkaline molecules such as these. CGE masses compared favorably to SDS-PAGE, even without making a correction for the contribution in mass from the oligosaccharide attached to the heavy chain. These results are similar to those published by Bennett et. al. [46] for bovine IgG and Kundu et. al. [51] for recombinant GB virus-C proteins. They also observed slightly larger masses for the protein when measured by CGE or SDS-PAGE.

8.7.5 Repeatability

Repeatability is evaluated by analyzing six independent samples at the same time using the same method. Tables III A and III B show the results of repeatability testing for migration time comparing CGE to SDS-PAGE. The acceptance criteria was less than 10% CV for the replicates. The data shows CGE migration times to have better repeatability than SDS-PAGE Rf values with typical CVs less than 1%.

8.7.6 Intermediate Precision

Intermediate precision is probably the best single parameter that can be used to estimate the variability associated with the method. For CGE, day-to-day, operator-to-operator, and instrument-to-instrument variables were examined. The results of those studies were combined into overall means and standard deviations, which are shown in Tables IV A and IV B. The data demonstrate that CGE has a lower overall CV for intermediate precision parameters when comparing migra-

Table II. Heavy and Light Chains Mass Comparisons. Synagis® Heavy and Light Chain Mass Comparison separated by SDS-PAGE, Capillary Gel Electrophoresis (CGE), and MALDI-TOF. SDS-PAGE and CGE mass calculations were determined by plotting the log MW versus migration time (min.) / distance migrated (Rf) of the molecular weight markers. The molecular size of heavy and light chain were determined using the linear regression curve.

Peak	Theoretical (Da)	CGE (Da)	MALDITOF (Da)	SDS-PAGE (Da)	MALDI/CGE (% difference)	SDS/CGE (% difference)
Heavy chain	51,000*	56,359	51,003	51,027	10.5	9.5
Deglycosylated heavy chain	49,464	48,657	49,235	48,196	1.2	1.0
Light chain	23,213	25,916	23,762	25,675	9.1	0.9

* Molecular weight of the heavy chain includes the average of attached oligosaccharide forms.

tion time to Rf. We found this to be the single most compelling advantage of CGE over SDS-PAGE in that the intermediate precision was approximately 3-fold better. The acceptance criteria for these validation studies was less than 10% CV.

8.7.7 Reproducibility

Reproducibility is a study of interlaboratory variability. This can typically be examined by performing an analysis of a sample panel in Development and QC laboratories. Migration time and peak purity results must be equivalent within 10%.

8.7.8 Robustness

The critical parameters most likely to affect the reliability of the method should be chosen for robustness testing. Example parameters for CGE include varying capillaries, applied voltage, salt concentration in the sample, number of consecutive samples, and injection time. Acceptance criteria should be based on Intermediate Precision studies. Based on the findings in the robustness studies, set operating ranges and revise the SOP appropriately.

8.7.9 Sensitivity

A commonly asked question regarding CGE is how does it compare in sensitivity to SDS-PAGE with Coomassie Blue-staining. Gump and Monnig [56] have reported using OPA, fluorescamine, and NDA to derivatize proteins with fluorescence detection to enhance sensitivity. Hunt and Nashabeh [53] have reported a method where proteins are derivatized using a fluorophore and the resulting derivatives are detected using laser-induced fluorescence resulting in sensitivity similar to silver staining. Harvey et. al. [57] have reported a subnanomolar detection limit using the noncovalent fluorescent dye, Sypro Red, with laser-induced fluorescence for detection.

We have attempted to demonstrate that UV detection in CGE is comparable to Coomassie Blue-staining in SDS-PAGE. In order to address this question, a study was designed based on two-fold dilution of the sample from the validated SOP condition until no more signal could be seen and comparing the signal-to-noise ratios between CGE and SDS-PAGE.

Table III A. Repeatability Precision Migration Times for Reducing CGE.

Peak	Mean Mt	Standard Deviation	% CV
92 kDa	15.56	0.059	0.38
Heavy chain	13.73	0.020	0.15
Light chain	11.35	0.003	0.03

$N = 6$; Mt = migration time of peak (minutes); (Adapted from [6]).

Table III B. Repeatability Precision Migration Times for Reducing SDS-PAGE.

Band	Mean Rf	Standard Deviation	% CV
92 kDa	0.483	0.008	1.7
Heavy chain	0.687	0.005	0.8
Light chain	0.900	0.006	0.7

$N = 6$; (Adapted from [6]).

Table IV A. Intermediate Precision Migration Times for Reducing CGE.

Peak	Mean migration time (minutes)	Standard Deviation	% CV
92 kDa	14.99	0.182	1.3
Heavy chain	13.25	0.172	1.2
Light chain	10.92	0.147	1.2

$N = 12$.

Table IV B. Intermediate Precision Rf values for Reducing SDS-PAGE.

Band	Mean Rf	Standard Deviation	% CV
92 kDa	0.472	0.021	4.0
Heavy chain	0.649	0.027	4.0
Light chain	0.861	0.028	3.0

$N = 12$; (Adapted from [6]).

The SDS-PAGE gel portion of this experiment is illustrated in Figure 4. As the protein is diluted, there is a linear decrease in the intensity of the stained heavy and light chain bands. A similar pattern was observed for CGE and a summary of the results is shown in Table V.

Synagis® was prepared according to the SDS-PAGE and CGE standard operating procedures. For SDS-PAGE, Synagis® was prepared as demonstrated in Figure 4. CGE samples were diluted 1:1 with sample buffer to yield a serial dilution of nine samples, and then analyzed by CE, containing 2.0 mg mL^{-1}, 1.0 mg mL^{-1}, 0.5 mg mL^{-1}, 0.25 mg mL^{-1}, 0.13 mg mL^{-1}, 0.063 mg mL^{-1}, and 0.031 mg mL^{-1} per sample of Synagis®. The signal-to-noise ratio was evaluated by both methods using the heavy chain peak signal versus the background noise. The HPCE ChemStation software provides system suitability performance tools to automatically perform the signal-to-noise ratio calculations. The SDS-PAGE method required analysis of the samples by image densitometry. Densitometry was performed by analyzing the electrophero-

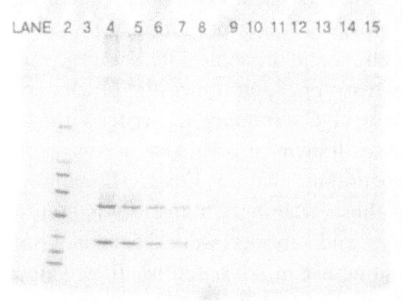

Figure 4. Reducing SDS-PAGE gel of Synagis® (Coomassie Blue-stained) for detection sensitivity. Synagis® was prepared with sample buffer containing 2% SDS with 5% β-mercaptoethanol, and heated to 80 °C for 10 minutes. Synagis® was then diluted 1:1 with sample buffer to yield a serial dilution of nine samples, and then loaded onto a 4–15% SDS-PAGE gel using 10 µL/lane, containing 3.0 µg, 1.5 µg, 0.75 µg, 0.38 µg, 0.19 µg, 0.094 µg, and 0.047 µg Synagis®. Lane assignments were: Bio-Rad high and low molecular weight markers (lane 2), and reduced Synagis® (lanes 4–12). Following electrophoresis at constant voltage, the gel was fixed and Coomassie Blue-stained. Molecular weight markers are: myosin (200,000 Da), β-galactosidase (116,250 Da), phosphorylase b (97,400 Da), bovine serum albumin (66,000 Da), ovalbumin (45,000 Da), carbonic anhydrase (31,000 Da), and soybean trypsin inhibitor (21,500 Da).

Table V. Comparison of SDS-PAGE to CGE signal-to-noise ratios.

Dilution Series	SDS-PAGE Signal-to-noise ratio	CGE Signal-to-noise ratio
No dilution	250	1686
2 × dilution	171	1044
4 × dilution	83	487
8 × dilution	44	176
16 × dilution	22	91
32 × dilution	9	58
64 × dilution	4	22
128 × dilution	No measurable signal	No measurable signal
256 × dilution	No measurable signal	No measurable signal

gram graphics of the banding profile for each sample loaded on the gel. The signal-to-noise ratio was determined by taking the graphical area of each heavy chain peak and dividing it by the average gel background. The background was determined by calculating the average graphical area background for all the blank sample lanes.

8.7.10 System Suitability

System suitability is perhaps the most overlooked but critical parameter in the validation of an analytical method. The selection of appropriate system suitability criteria can dictate the success or failure of a method in routine use. The criteria must be selected carefully based on a large body of data gathered during preclinical, clinical, and post-marketing manufacturing campaigns. It is important to apply statistically sound principles [58], sound scientific reasoning, and a measure of common sense in determining the criteria ranges that will be most predictive of a properly functioning method. For CGE methods, establish meaningful ranges for migration times and % purity. Once the method is in routine use in a QC setting, these values should be trended annually and tightened appropriately.

8.7.11 Validation Report

The validation report is the final documentation of the validation studies. It should summarize the findings of the validation study and define any OOS results or deviations from the preapproved protocol. In addition, it should provide complete, scientifically based explanations for any unanticipated results. The report should clearly define the limitations of the method and describe how the SOP will be modified based on the study findings. The report should clearly define references to

raw data to show full traceability to original results. Scientifically sound statistical analysis of data should be used, where possible, to draw conclusions. Finally, perform a complete audit of raw data prior to review of the report to assure data integrity. If in doubt about OOS results, report findings, or any other aspect of the validation study, consult with the Quality Assurance group.

8.7.12 Validation Tips and Traps

During the course of our validation studies for CGE, we came across many situations that complicated the data collection and report writing. We have gathered here a number of suggestions that we hope will allow others to avoid these pitfalls and streamline their validation studies.

- Carefully consider the validation strategy before performing any studies. Maybe the method does not require validation!
- Plan and execute validation pilot studies to determine acceptance criteria.
- Start as early as possible to gather data on meaningful system suitability criteria. Don't make these too tight, too early!
- Allow enough time for pilot studies and validation experiments. Unrealistic deadlines make for disaster in validation studies!
- Record all observations, deviations, and unusual events in a Field notebook (validation diary). Provide timely explanations for the events, if possible.
- Any values that fail the acceptance criteria should be handled as OOS. Perform a thorough investigation in collaboration with QA.
- Provide a standard repository for raw data. Make sure all reports are traceable to the raw data.
- Provide cross-references to any relevant computer and equipment validations.

- As part of the final report review, audit all raw data references.
- Explain why any validation parameters are not considered relevant.
- Explain any unusual or OOS results in the final report.
- Describe in the conclusions the limitations of the method and how the SOP will be revised.
- Maintain a log of system suitability failures.
- Maintain a constant feedback between the Development labs where the method originated and QC where it is routinely run.
- No validation study can predict every variable that could cause an assay problem. Consistent feedback and communications between Development and QC is the best source of information!

It is also vital to maintain vendor qualification programs. For example, the supplier for your CGE sieving solutions may decide to reformulate those solutions resulting in system suitability failures for new lots. Work with the vendor to assure that substantial quantities of prequalified lots can be manufactured, purchased, and placed on a stability monitoring program to verify the vendor recommendations (if any) for expiration dating. Future work will undoubtedly focus on further miniaturizing and automating CGE and related CE techniques [55].

8.8 Acknowledgements

The authors gratefully acknowledge the expert technical assistance of Amy Burch, Mike Scott (QC), and John Hope. In addition, the authors wish to thank Gail Folena-Wasserman and Tom Pritchett for their critical review of this manuscript.

8.9 References

[1] Landers, J.P. *Handbook of Capillary Electrophoresis*, CRC Press, Boca Raton, **1994**.
[2] Kuhn, R.; Hoffstetter-Kuhn, S. *Capillary Electrophoresis: Principles and Practice*, Springer-Verlag, Berlin, **1993**.
[3] Righetti, P.G. *Capillary Electrophoresis in Analytical Biotechnology*, CRC Press, Boca Raton, **1996**.
[4] Camilleri, P. *Capillary Electrophoresis Theory and Practice*, CRC Press, Boca Raton, **1993**.
[5] Khaledi, M.G. High-performance capillary electrophoresis: theory, techniques,

and applications. John Wiley and Sons, Inc., New York, **1998**.

[6] Bowen, S.H.; Schenerman, M.A. Replacing slab gel electrophoresis methods with capillary electrophoresis for quality control and stability testing. *BioPharm*, November, **1998**, 42–50.

[7] Jorgenson, J.W.; Lukacs, K.D. Capillary zone electrophoresis. *Science* **1983**, *222*, 266–272.

[8] Karger, B.L.; Cohen, A.S.; Guttman, A. High performance capillary electrophoresis in the biological sciences. *J. Chromatogr.* **1989**, *492*, 585–613.

[9] Pritchett, T.J. Quantitative analysis of monoclonal antibodies using three modes of capillary electrophoresis. *BioPharm* **1995**, *8(7)*, 38–45.

[10] Strege, M.A.; Lagu, A.L. Capillary electrophoresis of biotechnology-derived proteins. *Electrophoresis* **1997**, *18(12–13)*, 2343–2352.

[11] Hjerten, S. in *Electrophoresis '83*, Hirai H. (Editor), Walter de Gruyter, Berlin, **1983**, pp. 71–79.

[12] Cohen, A.S.; Karger, B.L. High performance sodium dodecyl sulfate polyacrylamide gel capillary electrophoresis of peptides and proteins. *J. Chromatogr.* **1987**, *397*, 409–417.

[13] Widhalm, A.; Schwer, C.; Blaas, D.; Kenndler, E. Capillary zone electrophoresis with a linear, non-cross-linked polyacrylamide gel: separation of proteins according to molecular mass. *J. Chromatogr.* **1991**, *549*, 446–451.

[14] Ganzler, K.; Greve, K.S.; Cohen, A.S.; Karger, B.L.; Guttman, A.; Cooke, N.C. High-performance capillary electrophoresis of SDS-protein complexes using UV-transparent polymer networks. *Anal. Chem.* **1992**, *64*, 2665–2671.

[15] Wu, D.; Regnier, F.E. (1992) Sodium dodecyl sulfate-capillary gel electrophoresis of proteins using non-cross-linked polyacrylamide. *J. Chromatogr.* **1992**, *608*, 349–356.

[16] Werner, W.; Demorest, D.; Wiktorowicz, J.E. Automated Ferguson analysis of glycoproteins by capillary electrophoresis using a replaceable sieving matrix. *Electrophoresis* **1993**, *14*, 759–763.

[17] Nakatani, M.; Shibukawa, A.; Nakagawa, T. Sodium dodecyl sulfate-polyacrylamide solution-filled capillary electrophoresis of proteins using stable linear polyacrylamide-coated capillary. *Biol. Pharm. Bull.* **1993**, *16*, 1185–1188.

[18] Guttman, A.; Nolan, J.; Cooke, N. Capillary sodium dodecyl sulfate gel electrophoresis of proteins. *J. Chromatogr.* **1993**, *632*, 171–175.

[19] Karim, M.R.; Janson, J.C.; Takagi, T. Size-dependent separation of proteins in the presence of sodium dodecyl sulfate and dextran in capillary electrophoresis: effect of molecular weight of dextran. *Electrophoresis* **1994**, *15*, 1531–1534.

[20] Nakatani, M.; Shibukawa, A.; Nakagawa, T.J. High-performance capillary electrophoresis of SDS-proteins using pullulan solution as separation matrix. *J. Chromatogr. A* **1994**, *672*, 213–218.

[21] Shieh, P.C.H.; Hoang, D.; Guttman, A.; Cooke, N. Capillary sodium dodecyl sulfate gel electrophoresis of proteins I. Reproducibility and stability. *J. Chromatogr. A* **1994**, *676*, 219–226.

[22] Guttman, A.; Shieh, P.; Lindahl, J.; Cooke, N. Capillary sodium dodecyl sulfate gel electrophoresis of proteins II. On the Ferguson method in polyethylene oxide gels. *J. Chromatogr. A* **1994**, *676*, 227–231.

[23] Guttman, A.; Nolan, J. Comparison of the separation of proteins by sodium dodecyl sulfate-slab gel electrophoresis and capillary sodium dodecyl sulfate-gel electrophoresis. *Anal. Biochem.* **1994**, *221*, 285.

[24] Bennedek, K.; Thiede, S. High-performance capillary electrophoresis of proteins using sodium dodecyl sulfate-poly(ethylene oxide). *J. Chromatogr. A* **1994**, *676*, 209–217.

[25] Simò-Alfonso, E.; Conti, M.; Gelfi, C.; Righetti, P.G. Sodium dodecyl sulfate capillary electrophoresis of proteins in entangled solutions of poly(vinyl alcohol). *J. Chromatogr. A* **1995**, *689*, 85–96.

[26] Nakatani, M.; Shibukawa, A.; Nakagawa, T. Effect of temperature and viscosity of sieving medium on electrophoretic behavior of sodium dodecyl sulfate-proteins on capillary electrophoresis in presence of pullulan. *Electrophoresis* **1996**, *17(7)*, 1210–1213.

[27] Guttman, A. Capillary sodium dodecyl sulfate-gel electrophoresis of proteins. *Electrophoresis* **1996**, *17*, 1333–1341.

[28] Takagi, T. Capillary electrophoresis in presence of sodium dodecyl sulfate and a sieving medium. *Electrophoresis* **1997**, *18*, 2239–2242.

[29] "CE-SDS Protein Kit", Bio-Rad US Bulletin **1874**.

[30] Laemmli, U.K. *Nature* (London) **1970**, *277*, 680.

[31] Wielgos, T.; Turner, P.; Havel, K. Validation of analytical capillary electrophoresis methods for use in a regulated environment. *J. Cap. Elec.* **1997**, *4:6*, 273–278.

[32] Swartz, M.E.; Krull, I.S. *Analytical method development and validation*, Marcel Dekker, New York, **1997**.

[33] International Conference on Harmonization Topic Q2B: Guideline on the validation of analytical procedures: methodology, Federal Register **1997**, *62(96)*, 27464–27467.

[34] International conference on harmonization: guideline on validation of analytical procedures: definitions and terminology. Federal Register **1995**, *60(40)*, 11260–11262.

[35] United States Pharmacopeia **2000**, 24/NF19, Validation of compendial methods <1225>, United States Pharmacopeial Convention, Rockville, MD.

[36] Center for Drug Evaluation and Research guidance document **1994** Validation of chromatographic methods, U.S. Food and Drug Administration.

[37] DeSain, C.; Sutton, C.V. Test method development and validation. *BioPharm*, **1997**, April, 60–63.

[38] Hokanson, G.C. A life cycle approach to the validation of analytical methods during pharmaceutical product development, Part 1: the initial method validation process. *Pharmaceutical Technology*, September, **1994**, 118–130.

[39] Gold, D.H. Validation: why, what, when, how much. *PDA Journal of Pharmaceutical Science and Technology* **1996**, *50(1)*, 55–60.

[40] Pritchett, T.J. Analytical development for biopharmaceutical quality control. *BioPharm* **1996**, *9*, 34–39.

[41] McEntire, J. Selection and validation of analytical techniques. *BioPharm* **1994**, *7*, 68–80.

[42] Altria, K.; Rudd, D. An overview of method validation and system suitability aspects in capillary electrophoresis. *Chromatographia* **1995**, *41*, 325–331.

[43] Kunkel, A.; Dengenhardt, M.; Schrim, B.; Watzig, H. Performance of instruments and aspects of methodology and validation in quantitative capillary electrophoresis: an update. *J. Chromatogr. A* **1997**, *768*, 17–27.

[44] Silverman, C.; Komar, M.; Shields, K.; Diegnan, G.; Adamovics, J. Separation of the isoforms of a monoclonal antibody by gel isoelectric focusing, high performance liquid chromatography and capillary isoelectric focusing. *J. Liq. Chromatogr.* **1992**, *15*, 207.

[45] Costello, M.A.; Woititz, C.; DeFeo, J.; Stremlo, D.; Wen, L.-F.L.; Palling, D.J.; Iqbal, K.; Guzman, N.A. (1992) Characterization of humanized anti-TAC monoclonal antibody by traditional separation techniques and capillary electrophoresis. *J. Liq. Chromatogr.* **1992**, *15*, 1081.

[46] Bennett, L.E.; Charman, W.N.; Williams, D.B.; Charman, S.A. (1994) Analysis of bovine immunoglobulin G by capillary gel electrophoresis. *J. Pharm. Biomed. Anal.* **1994**, *12*, 1103–1108.

[47] Liu, J.; Abid, S.; Lee, M.S. Analysis of monoclonal antibody chimeric BR96-doxorubicin immunoconjugate by sodium dodecyl sulfate-capillary electrophoresis with ultraviolet and laser-induced fluorescence detection. *Anal. Biochem.* **1995**, *229*, 221–228.

[48] Moorhouse, K.G.; Eusebio, C.A.; Hunt, G.; Chen, A.B. Rapid one-step capillary isoelectric focusing method to monitor charged glycoforms of recombinant human tissue-type plasminogen activator. *J. Chromatogr. A* **1995**, *717*, 61.

[49] Hunt, G.; Moorhouse, K.G.; Chen, A.B. Capillary isoelectric focusing and sodium dodecyl sulfate-capillary electrophoresis of recombinant humanized monoclonal antibody HER2. *J. Chromatogr. A* **1996**, *744(1–2)*, 295–301.

[50] Moorhouse, K.G.; Rickel, C.A.; Chen, A.B. Electrophoretic separation of recombinant tissue-type plasminogen activator glycoforms: validation issues for capillary isoelectric focusing methods. *Electrophoresis* **1996**, *17*, 423.

[51] Kundu, S.; Fenters, C.; Lopez, M.; Varma, A.; Brackett, J.; Kuemmerle, S.; Hunt, J.C. (1997) Capillary electrophoresis for purity estimation and in-process testing of recombinant GB virus-C proteins. *J. Capillary Electrophoresis* **1997**, *4(1)*, 7–13.

[52] Buchacher, A.; Schulz, P.; Choromanski, J.; Schwinn, H.; Josic, D. High performance capillary electrophoresis for in-process control in the production of antithrombin III and clotting factor IX. *J. Chromatogr. A* **1998**, *802(2)*, 355–366.

Original

[53] Hunt, G.; Nashabeh, W. Capillary electrophoresis sodium dodecyl sulfate nongel sieving analysis of a therapeutic recombinant monoclonal antibody: a biotechnology perspective. *Anal. Chem.* **1999**, *71*, 2390–2397.

[54] Stocks, J.; Miller, N.E. Analysis of apolipoproteins and lipoproteins by capillary electrophoresis. *Electrophoresis* **1999**, *20*, 2118–2123.

[55] Yao, S.; Anex, D.S.; Caldwell, W.B.; Arnold, D.W.; Smith, K.B.; Schultz, P.G. SDS capillary gel electrophoresis of proteins in microfabricated channels. *Proc. Natl. Acad. Sci.* (USA) **1999**, *96*, 5372–5377.

[56] Gump, E.L.; Monnig, C.A. Pre-column derivatization of proteins to enhance detection sensitivity for sodium dodecyl sulfate non-gel sieving capillary electrophoresis. *J. Chromatogr. A* **1995**, *715*, 167–177.

[57] Harvey, M.D.; Bandilla, D.; Banks, P.R. Subnanomolar detection limit for sodium dodecyl sulfate-capillary electrophoresis using a fluorogenic, noncovalent dye. *Electrophoresis* **1998**, *19(12)*, 2169–2174.

[58] Weed, D.H. A statistically integrated approach to analytical method validation. *Pharmaceutical Technology*, October, **1999**.

Analysis of Protein Therapeutics by Capillary Electrophoresis

S. Ma / W. Nashabeh

Department of Analytical Chemistry, Department of Quality Control Analytical Technologies, Genentech. Inc., USA

9.1 Introduction

The development, manufacture and quality control of protein therapeutics from recombinant DNA technology have required state-of-the-art analytical methodologies to elucidate protein structure and to assure the homogeneity, purity, and molecular stability of the products. In this chapter, we intend to demonstrate that capillary electrophoresis (CE) can play an important role in the analysis of protein therapeutics, and significantly enrich the arsenal of analytical methodologies by providing a complementary technique in the evaluation of the physcio-chemical attributes of proteins in terms of charge, hydrophobicity, or size. In particular, CE has been extremely useful in replacing the traditional slab gel electrophoresis techniques, such as CE-SDS as a replacement of SDS-PAGE and cIEF as a replacement of IEF gels. Selected applications that illustrate the advantages of CE as a replacement to gel electrophoresis will be shown at various steps of the manufacture of recombinant therapeutics including formulation development, large scale recovery process development, lot release testing and stability monitoring. In addition, CZE is shown to provide an orthogonal separation technique to liquid chromatography, particularly in assessing the charge heterogeneity of proteins. Finally, the high resolving power and sensitivity of CE with laser-induced fluorescence detection is illustrated in the analyses of carbohydrate moieties on proteins, including both oligosaccharides and monosaccharide

analyses. The majority of the applications described here have been validated according to the guidelines of the International Committee on Harmonization, and are currently being used in routine lot release testing and stability monitoring of selected Genentech marketed protein therapeutics.

This chapter is divided into four main sections based on the distinct modes of capillary electrophoresis discussed: (i) capillary electrophoresis-sodium dodecylsulfate, (ii) capillary isoelectric focusing, (iii) capillary zone electrophoresis, and (iv) carbohydrate analysis. Each section contains a brief description of the theory and key fundamentals of the CE methodology of interest, followed by selected applications currently practiced at Genentech. A detailed methods section is provided at the end of the chapter for the convenience of the readers.

9.2 Capillary Electrophoresis – Sodium Dodecylsulfate (CE-SDS)

Among the various size-based separation methods, high-performance size-exclusion chromatography (HPSEC) provides accurate and reliable separations of proteins in their aggregate or monomeric forms, but is generally limited in its ability to separate more closely related size variants, such as light/heavy chain fragments observed in recombinant monoclonal antibodies. Sodium dodecylsulfate-polyacry-

lamide gel electrophoresis (SDS-PAGE), on the other hand, typically provides superior separations of a denatured protein and has been used for over 25 years as the primary method of choice for size-based protein separations [1]. In SDS-PAGE, proteins are bound to SDS on a relatively constant weight basis, about 1.4 gram of SDS per gram of protein [2]. Consequently, the SDS-protein complexes all share a similar random coil shape and molecular charge per unit mass. Therefore, the free solution mobilities of the SDS-protein complexes are practically identical [3]. When a mixture of SDS-saturated proteins is electrophoresed in a sieving medium, separation is achieved based on the molecular size or hydrodynamic radius. Furthermore, the electrophoretic mobility of the complex is proportional to the log of the apparent molecular size which is proportional to the molecular weight of the polypeptide backbone chain [4].

Besides being a tool to estimate the apparent molecular weight of proteins, SDS-PAGE has been widely used to monitor impurities and confirm the consistency of manufacture of biologics. Depending upon the intended use of SDS-PAGE analyses and the sensitivity required, visualization of the separated proteins is generally accomplished by staining with either Coomassie Brilliant Blue R-250 [5] or the more sensitive silver-stain dyes [6]. Although Coomassie Brilliant Blue is widely used for semi-quantitative evaluation of the relative intensity of the major protein bands, its limit of detection is often considered inadequate for low level

0009-5893/00/02   75-14 $ 03.00/0   © 2001 Friedr. Vieweg & Sohn Verlagsgesellschaft mbH

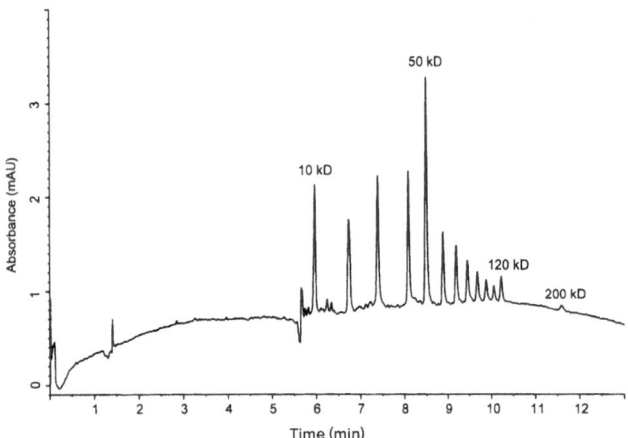

Figure 1. CE-SDS/UV of protein molecular weight ladder. The standard masses range from 10 to 120 kD in 10 kD increments and an additional 200 kD protein. Capillary: fused silica capillary (50 μm × 27 cm); buffer: CE-SDS run buffer (BioRad); voltage: 15 kV, reversed polarity; capillary temperature, 20 °C; injection: electrokinetic at 10 kV for 15 sec; detection, UV at 220 nm.

impurity detection. Use of the silver-stain method significantly improves the LOD to 10 ng mL^{-1} as determined by analysis of bovine serum albumin (BSA) as a sensitivity marker [7], which is about 100 fold more sensitive than that of the Coomassie Blue. Nevertheless, silver-stained SDS-PAGE approach suffers from several limitations: labor intensive (gel preparation, separation, staining, destaining), use of toxic reagents, and high intra- and intergel effective mobility and staining variability.

While CE utilizes the same separation principles as gel electrophoresis, it offers the added advantages of enhanced precision due to automation, speed as the result of higher field strength, and, most importantly, on-line detection to replace the laborious staining and destaining procedures. For the past two decades, there has been increasing activity in evaluating CE as an automated and instrumental approach to classical electrophoresis. The initial work on CE separations with sieving matrices was performed with crosslinked polyacrylamide gel-filled capillaries [8–11]. Although the separation of oligonucleotides and proteins, to a less extent, was achieved, problems associated with the reproducibility of gel fabrication and limited flexibility of the rigid separation matrix have hampered the widespread use of gel-filled capillaries. Over the next few years, linear (non-crosslinked) polymer networks became the focus of the development. In 1989, Zhu and his coworkers [12] successfully incorporated solutions of large linear polymers, such as methylcellulose or polyethylene glycol as buffer additives, to provide a

molecular sieving effect that facilitated the separation of DNA fragments and protein size variants, e. g., monomer, dimer, and trimer forms of BSA. The use of polymer solutions allows for the replacement of the sieving matrix after each CE analysis, therefore enhancing the overall precision and robustness. The successful employment of the linear polymer networks in the DNA sequencing field has made an enormous contribution in the successful mapping of the human genome. For CE-SDS separations of proteins, a number of polymers have also been reported in the literature, including linear polyacrylamide solutions, dextran [13], pullulan [14], and linear polyethylene oxide [15]. The replaceable polymers were effective in resolving SDS-protein complexes on the basis of their size, with the ability to provide rapid and accurate estimates of molecular weight and size variant analyses. Further details on the principles and applications of this technique can be found in the review articles by Guttman [16] and Takagi [17].

In this report, the applications of CE-SDS for recombinant protein analysis will be presented. Similar to SDS-PAGE, the applications will be categorized based upon the modes of detection: UV absorbance and lasser induced fluorescence (LIF). The use of UV absorbance detection is by far the simplest mode of CE-SDS because there is no sample preparation beyond the SDS-protein binding step required prior to analysis. CE-SDS with UV detection provides equivalent sensitivity to Coomassie-blue staining. Hence, it can be used as a versatile tool for molecular weight estimation and accurate

quantitative monitoring of protein variants present at levels as low as 0.5% in terms of their relative peak areas with respect to the main product [18]. For applications that require higher sensitivity, precolumn derivatization of the proteins with a fluorophore prior to CE-SDS analysis using laser-induced fluorescence detection has been shown to provide at least 100-fold increase in detection sensitivity at 10 ng mL^{-1} using a recombinant human monoclonal antibody (rhuMab). CE-SDS with LIF detection provides equivalent sensitivity to silver-stained SDS-PAGE and can be used for detection and analysis of low-level protein variants and contaminants with peak area representation of approximately 0.001 – 0.005% relative to main product [18]. However, a practical limitation to this approach is the additional sample manipulation involving the protein derivatization. Another approach using laser-induced fluorescence detection based on the native protein fluorescence from tryptophan or tyrosine residues. This approach is expected to provide enhanced sensitivity over UV detection is without additional sample manipulation. The feasibility of this approach is currently under investigation in our laboratories. In the following sections, the applications of CE-SDS for protein therapeutics using both UV absorbance and LIF detection modes will be discussed.

9.2.1 CE-SDS with UV Detection

The electropherogram of a protein molecular mass ladder with masses ranging from 10 to 120 kD in 10 kD increments and a 200 kD standard separated by CE-SDS is shown in Fig. 1. The separation is achieved using a commercial UV-transparent hydrophilic polymer as the sieving matrix. A plot of the logarithmic molecular mass of the protein standards versus their electrophoretic mobility exhibited good linearity with a correlation coefficient greater than 0.99 in the mass range studied here. With on-column UV detection, estimation of molecular mass as well as accurate quantitation of relative molecular mass distribution can be readily obtained without any labor-intensive staining/destaining procedures required in SDS-PAGE. Although the majority of the CE-SDS applications reported to date demonstrated the technique's ability to determine protein's intrinsic property, such as apparent molecular weight, our pri-

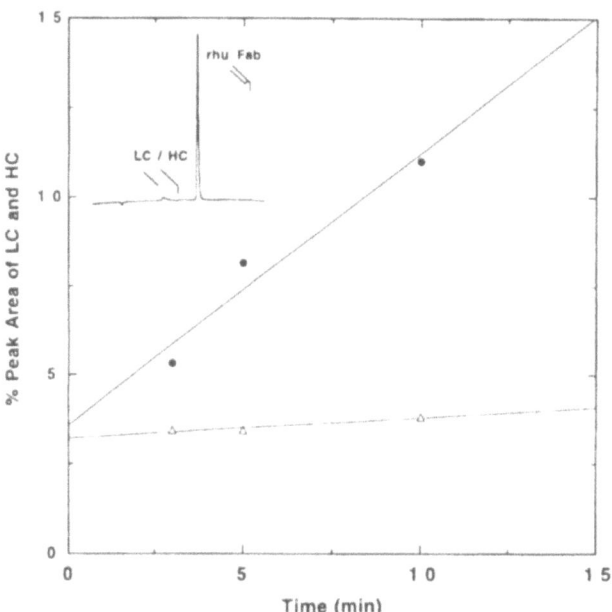

Figure 2. Plots of the percent peak area of the light/heavy chain in a rhu-Fab versus the time of SDS-protein binding at different temperatures: (circle) 90 °C and (triangle) 60 °C. Capillary: fused silica capillary (50 µm × 27 cm); buffer: CE-SDS run buffer; voltage: 15 kV, reversed polarity; capillary temperature, 20 °C; injection: hydrodynamically at 40 psi · sec; detection, UV at 220 nm.

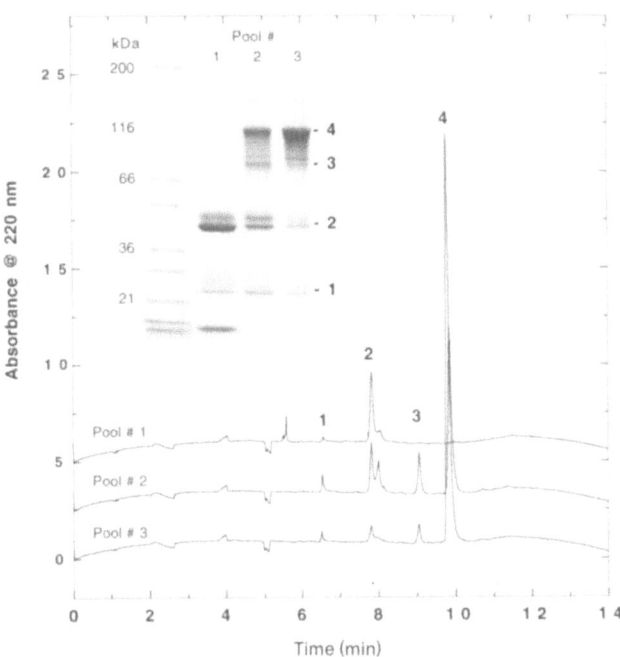

Figure 3. Electropherograms of three different pools obtained during a purification step for the manufacture of a rhu F(ab')₂. The corresponding SDS-PAGE of the same pools is also shown as insert here. Peaks and bands tentatively identified as: (1) free light or heavy chain; (2) Fab, light chain homodimer or heavy chain homodimer; (3) heavy/heavy/light chain trimer; (4) intact rhu F(ab')₂. Capillary: fused silica capillary (50 µm × 27 cm); buffer: CE-SDS run buffer; voltage: 15 kV, reversed polarity; capillary temperature, 20 °C; injection: hydrodynamically at 40 psi · sec; detection, UV at 220 nm.

mary interest in CE-SDS is the molecular mass distribution. It should be noted this aim is fundamentally different from those found in the literature. The methodology is herein intended to provide a size-based fingerprint of a given product in terms of its related size variants such as fragments, aggregates, and/or any non-product related impurities, such as host cell proteins. Consequently, when compared to a well-characterized reference material, it provides information on the product purity as well as the consistency of manufacture.

Sample Preparation. A critical consideration in using CE-SDS to asses size distribution of recombinant protein therapeutics is a thorough understanding of the impact of sample preparation, e. g., SDS-protein binding step on the analysis. It's been previously reported [18] that the traditional practice of achieving SDS-protein binding at elevated temperatures (90 °C or higher) often causes thermally induced fragmentation and/or aggregation. This is especially true for multiple disulfide-linked polypeptides, such as recombinant monoclonal antibodies. The artifacts not only significantly alter the true representation of the protein but also increase the variability of the assay. For example, the rhuMab fragments can increase from an

average of 4% at 1 min to 21% after 10 min of heating at 90 °C when analyzed under non-reducing conditions [18]. The main mechanism of thermal degradation appears to occur initially through inter-chain disulfide bond rupture between the light and heavy chains followed by the disulfide bonds holding the two heavy chains. Similar behavior was observed for a recombinant human monoclonal Fab antibody (rhuFab) wherein only one light chain is disulfide linked to one heavy chain. A typical electropherogram for the CE-SDS of a rhuFab is shown in Fig. 2. The percent peak area of the free light or heavy chain (LC/HC) obtained at 90° and 60 °C was determined and then plotted against the heating time in Fig. 2. It is seen from the figure that the free LC/HC fragments increased significantly from about 5% to 12% at 90 °C for 5 and 15 minutes, respectively. Upon lowering the temperature to 60 °C, the LC/HC fragments remained fairly constant at about 3–4% over the same time period. Furthermore, by extrapolating the two temperature curves to time zero, a "true" representation of the amount of free LC/HC in the sample was determined to be 3.5%. This example demonstrates the advantage of CE-SDS in providing accurate quantita-

tion of relative protein distribution with the on-column detection. In addition, it also serves as a caveat that the SDS-protein binding step should be carefully optimized, in terms of both time and temperature, in order to maintain an accurate and reproducible sample preparation.

Recovery Process Development of a rhuF(ab')₂. SDS-PAGE has been widely used in the recovery process development to verify the purity of the product. During the development of a chromatographic step, three fractions were pooled from the elution step and analyzed by SDS-PAGE with Coomassie staining (Fig. 3). Four main bands were detected at various levels. They were identified based on the molecular weight standards and Edman sequencing data to be the one chain (LC), two chains (Fab, LL or HL), three chains (HHL), and the intact F(ab')₂. The inherent variability of the technique in terms of the resolution and staining, particularly, in the region for band 3 and 4, made it difficult to assess the purity in a quantitative manner and hence to obtain pooling criteria. In order to better quantitate the purity, the same pools were analyzed by CE-SDS with UV detection and results are shown in Fig. 3. It is clear that the overall peak pattern matches the band pattern.

Original

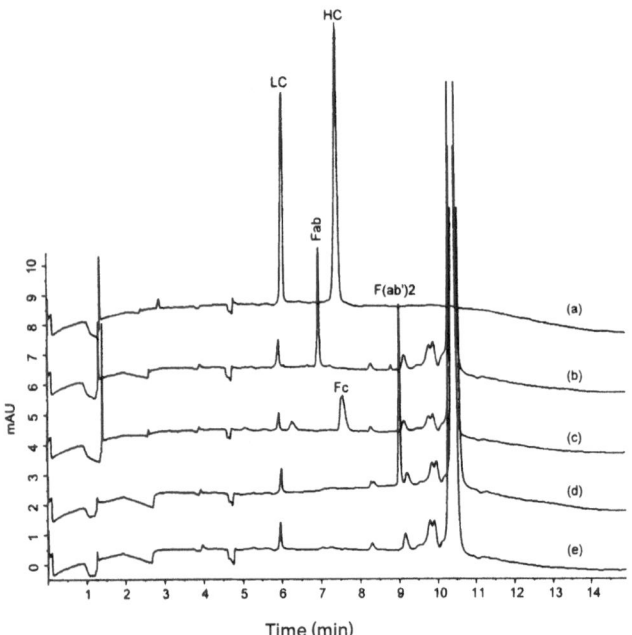

Figure 4. CE-SDS/UV of common monoclonal antibody fragments generated either chemically or proteolytically: (**a**) light and heavy chain generated from DTT reduction of a Mab; (**b**) Fab generated by papain digestion spiked into a Mab; (**c**) Fc generated by papain digestion spiked into a Mab; (**d**) F(ab')₂ generated by pepsin digestion spiked into a Mab; (**e**) intact Mab. Capillary: fused silica capillary (50 μm × 27 cm); buffer: CE-SDS run buffer (BioRad); voltage: 15 kV, reversed polarity; capillary temperature, 20 °C; injection: electrokinetic at 10 kV for 15 sec; detection, UV at 220 nm. (From [18]. With permission.)

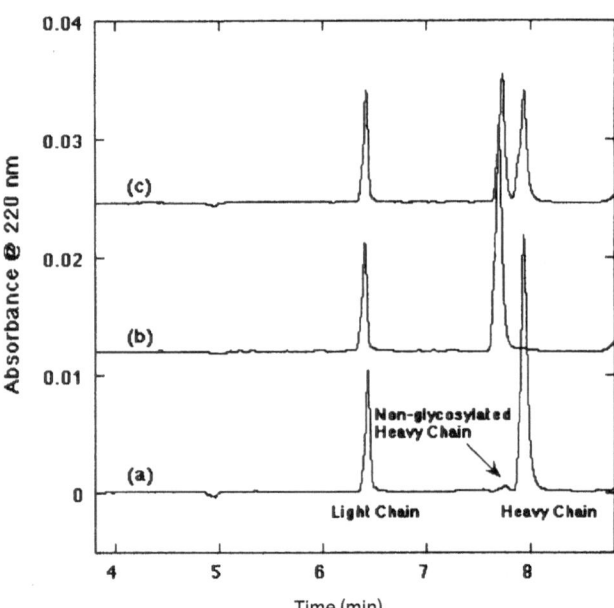

Figure 5. Determination of glycosylation occupancy at the Fc region of rhuMabs by CE-SDS/UV. (**a**) a reduced rhuMab; (**b**) after PNGase F digestion; (**c**) comix of a and b. Capillary: fused silica capillary (50 μm × 27 cm); buffer: CE-SDS run buffer (Bio-Rad); voltage: 15 kV, reversed polarity; capillary temperature, 20 °C; injection: hydrodynamically at 40 psi · sec; detection, UV at 220 nm.

Again, the data suggests that comparable resolution can be achieved in CE-SDS with a uniform polymer solution instead of a gradient gel largely due to the improved efficiency. In addition, the purity of each pool can be readily determined and pooling criteria unambiguously set.

Separation of rhuMab and its common fragments. A typical example of CE-SDS analysis of a rhuMab and its fragments is illustrated in Fig. 4. Under nonreducing conditions (Fig. 4e), the CE electropherogram depicts a characteristic fingerprint of the various size-variants observed in a rhuMab preparation due to the different combinations of light and heavy chains. The identity of the peaks in the CE profile were tentatively assigned [18] based on a comparison with the bands in SDS-PAGE, whose identities were obtained through sequential Edman degradation. Upon chemical reduction with either DTT or BME, the protein is reduced to one light chain and one heavy chain as shown in Fig. 4a. Several commonly found proteolytic fragments, i. e., F(ab')₂, Fab or Fc fragments, were generated via digestion with pepsin or papain and then spiked into the nonreduced rhuMab preparation (Fig. 4b–4d). It is seen from the

figure that all these fragments were clearly resolved from each other demonstrating the resolving power of CE-SDS with a polymer solution as the sieving matrix. Even though the heavy chain, Fab and Fc all have about the same molecular weight of 50 kD, these three species are still separated from each other with the migration order of Fab, HC and Fc. This anomalous migration is attributable to the increase in hydrodynamic sizes as consequences of glycosylation of the rhuMab molecules in each of the heavy chain (HC), through the attachment of a neutral *N*-linked, complex, asialo-biantennary glycan with a core fucose. As such, the heavy chain containing a single glycosylation site exhibited a larger size and thus higher apparent mass than the Fab fragment devoid of the glycosylation site. On the other hand, HC migrated faster than the Fc fragment as it contained a single glycosylation site compared with two glycosylation sites in the Fc fragment. This superior resolution over the traditional SDS-PAGE provides the foundation to further explore the scope of CE-SDS, particularly, the application illsutrated below.

Determination of glycosylation occupancy. It is well known that the carbohy-

drates play a key role in the therapeutic use of monoclonal antibodies [19, 20]. The extent of glycosylation through an *N*-linked, complex, asialo-biantennary structure is therefore routinely monitored. Traditionally, the data is extracted from a peptide map wherein the methodology requires relatively large amount of material and is often time consuming. CE-SDS, on the other hand, requires minute amounts of material and the sample preparation only involves the SDS-protein binding with reduction to simplify the profile to a single heavy chain with or without glycosylation at the site. A typically CE profile of a reduced rhuMab is shown in Fig. 5a. To demonstarte the effects of glycosylation on apparent migration in CE-SDS, a rhuMab molecule was deglycosylated using PNGase F, an endoglycosidase that hydrolyzes the β-aspartylglucosamine bond between the asparagine residue and the innermost *N*-acetylglucosamine of the glycan. The results conform the earlier observation that upon removal of the carbohydrate, the non-glycosylated heavy chain (Fig. 5b) shifted to an earlier migration time due to its decreased hydrodynamic size and is resolved from the glycosylated heavy chain at the baseline (Fig. 5c). The limit of detection for this nonglycosylated speices is about 0.5% (UV detection limits mentioned earlier). Although the examples shown here involve only one glycosylation site, the methodology applies to proteins with multiple glycosylation sites provided sufficient resolution can be achieved in CE-SDS.

Extent of PEGylation of a rhuFab. There has been increasing interest in exploring the therapeutic use of antibody fragments, such as a Fab, instead of their full length counterparts. In order to extend the circulating half-life of Fab molecules, polyethylene glycols (PEG) are used to conjugate proteins and the extent of the pegylation needs to be monitored. Over the years, various analytical tools, i.e., HPLC, MS and NMR, have been used to assess the yield. Unfortunately, due to the heterogenous nature of the PEG as well as the inherent complexity of recombinant protein preparations, the results are often difficult to interpret. CE-SDS, on the other hand, separates species based on the different sizes or mass and therefore is expected to provide an alternative to measure the pegylation yield. A rhuFab molecule was conjugated at a single site with either a 20 kD PEG or 40 kD PEG and analyzed by CE-SDS with the results shown in Fig. 6. It is clear that separation is achieved between these three speices and the pegylation yield can be easily determined to be greater than 95% in these two cases. Moreover, the complexity due to the heterogenous nature of the PEG seen in other technique is minimized as clearly demonstrated by the sharpness of the peaks obtained here. Although this example is for a single peglyation site, the data strongly suggest that this approach can be extented to evaluate the pegylation extent at multiple sites.

The examples shown here demonstrate the advantages of CE-SDS with UV detection over the SDS-PAGE in terms of its speed, automation, and ease of on-line detection. The LOD of UV detection has been reported to comparable to that is routinely achieved with Coomassie Blue staining. Attempts to increase absorption sensitivity through combinations of larger inner-diameter capillaries and /or larger injection plugs resulted in limited success at the expense of loss in resolution. Despite its limitation in sensitivity, UV detection is as useful for monitoring the consistency of manufacture as it is often done using SDS-PAGE visualized with Coomassie Blue staining. Nevertheless, an al-

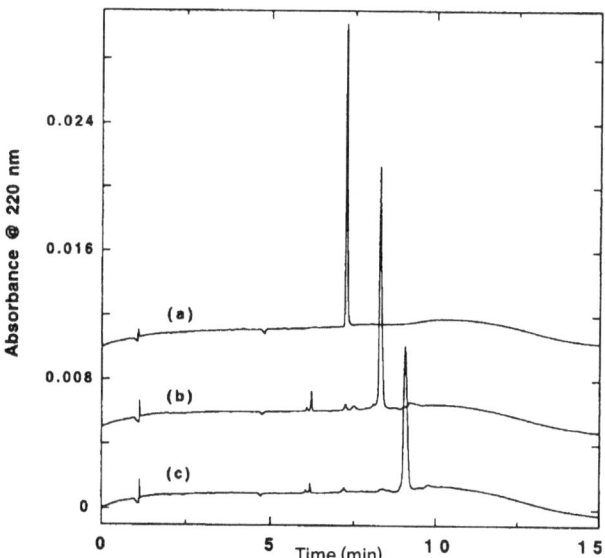

Figure 6. Extent of PEGlylation of a rhuFab determined by CE-SDS/UV. (a) a rhuFab; (b) rhuFab with 20 kD PEG; (c) rhuFab with 40 kD PEG. Capillary: fused silica capillary (50 μm × 27 cm); buffer: CE-SDS run buffer (BioRad); voltage: 15 kV, reversed polarity; capillary temperature, 20 °C; injection: hydrodynamically at 40 psi · sec; detection, UV at 220 nm.

ternative detection scheme, such as the laser-induced fluorescence (LIF) detection, is needed for improved sensitivity.

9.2.2 CE-SDS with Laser-Induced Fluorescence (LIF) Detection

The abundance of the lysine residues in most proteins provides ideal reactive sites for precolumn labeling because their aliphatic epsilon-amines are reasonably good nucleophiles above pH 8.0 [21]. The aliphatic amines, in addition to the N-terminal amino group, react with succinimidyl esters with high selectivity in the pH range 8–9, resulting in the formation of stable amide bonds [22]. Among the various fluorophores that can be readily excited by the Argon-ion laser (readily available in commercial CE instruments), a highly purified 5-isomer succinimidyl ester derivative of carboxytetramethylrhodamine (5-TAMRA.SE) is selected due to its intrinsic photostability and superior spectral properties. In particular, its insensitivity to pH changes between 4 and 10 provides an important advantage over fluoresceins for CE applications.

CE-SDS/LIF of a rhuMab. A typical example of CE-SDS analysis of a rhuMab derivatized with 5-TAMRA.SE is illustrated in Figs. 7a and b under reducing and non-reducing conditions, respectively [18]. Also shown in the figure are results

from the silver-stained SDS-PAGE analysis of the same sample. The data demonstrate good agreement in the number and relative intensity of the slab-gel bands and the peaks in the CE profile. The improved sensitivity is clearly illustrated by the presence of the minor peaks, which were previously undetected with UV detection (Figs. 4a and 4e). It should be pointed out that the separation between the rhuMab species is not significantly affected by the addition of the fluorophore, as evident in the identical electrophoretic profiles in Figs. 4 and 7 (nonlabeled vs labeled samples). Despite the heterogeneous nature of the labeling reaction, no dye-related artifact peaks, multiple peak formation from single analytes, or loss of resolution were observed. In addition, the LOD for the monomer peak, under optimized conditions, in this case was previously reported to be about $10 \, \text{ng mL}^{-1}$, compares well with silver-stained SDS-PAGE with 140-fold increase over UV detection run under the same conditions [18].

Low level impurity detection. Since one of the primary uses of SDS-PAGE is the detection of low level impurities, we evaluated CE-SDS with LIF detection for the purpose of detecting protein impurities that are non-product related. Such impurities can result from host cell proteins or cross contamination from other products manufactured in the same facility. A rhuMab was spiked with 4 know proteins at a

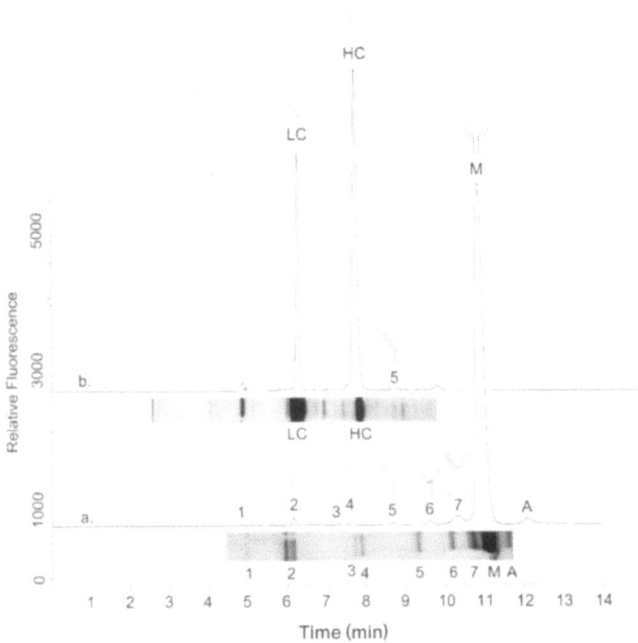

Figure 7. CE-SDS with laser induced fluorescence detection of a rhuMab derivatized with 5TAMRA.SE under different conditions: (**a**) nonreduced; (**b**) reduced. Capillary: fused silica capillary (50 μm × 27 cm); buffer: CE-SDS run buffer (Bio-Rad); voltage: 15 kV, reversed polarity; capillary temperature, 20 °C; injection: electrokinetic at 10 kV for 15 sec; detection, LIF with a 3.5 mW argon ion laser, 488 nm excitation/560 ± 20 nm emission. (From [18]. With permission.)

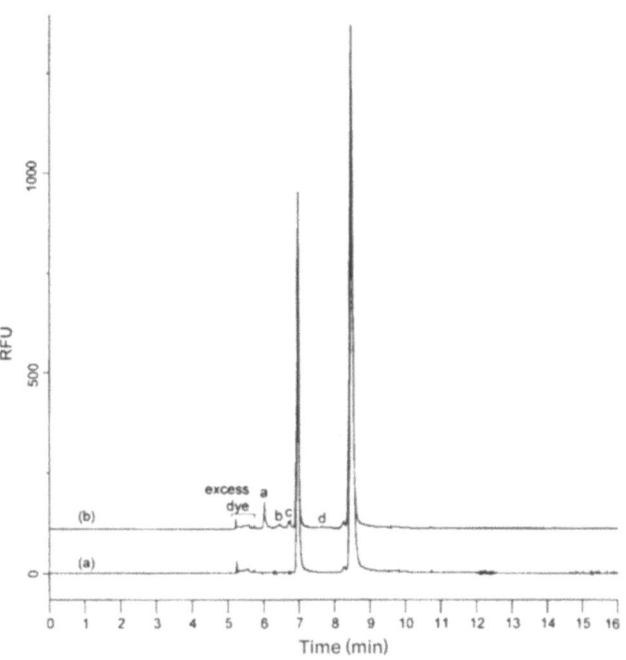

Figure 8. Monitor low level process impurities by CE-SDS/LIF. Samples: (**a**) rhuMab; (**b**) rhuMab spiked with four known proteins, each at a concentration of 0.5% (w/w) or 5000 ppm. Capillary: fused silica capillary (50 μm × 27 cm); buffer: CE-SDS run buffer (Bio-Rad); voltage: 15 kV, reversed polarity; capillary temperature, 20 °C; injection: electrokinetic at 10 kV for 15 sec; detection, LIF with a 3.5 mW argon ion laser, 488 nm excitation/560 ± 20 nm emission.

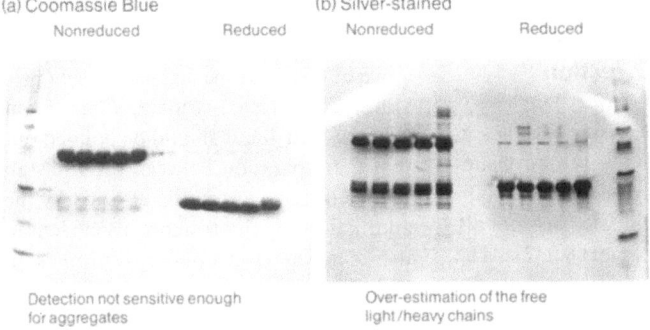

Figure 9. SDS-PAGE of a rhuFab in different formulations.

level of 0.5% (w/w) or 5000 ppm and then labeled with 5-TAMRA.SE prior to the CE analysis. The profiles of the reduced rhuMab alone and that with the protein spikes are shown in Fig. 8. The signal-to-noise ratios of the proteins spiked at 5000 ppm ranged from 957 (small 13 kD basic protein) to 38 (heavily glycosylated acidic protein). Using such a diverse protein spike mixture, the mean signal-to-noise ratio was 350, indicating that the sensitivity of the method in terms of impurities detection is about 45 ppm (LOD defined as a signal-to-noise ratio of 3). The example shown here demonstrated the efficiency of the labeling scheme in tagging such low level protein concentrations and

more importantly the potential of detecting low level impurity.

Formulation Development. SDS-PAGE is often used in formulation development of protein therapeutics and in monitoring stability. An example of SDS-PAGE of a rhuFab is shown in Fig. 9 where the gels are either stained with Coomassie Blue (Fig. 9a) or silver (Fig. 9b). The rhuFab in five different formulations were analyzed under both non-reduced and reduced conditions, where free light or heavy chain, intact Fab and aggregates were expected to exist. By comparing the two gels, it is clear that neither staining procedure can provide accurate quantitation of these three species. In the silver-stained gels, all the

species are detected but the amount of free light or heavy chain are clearly over estimated. On the other hand, with Coomassie Blue staining, the sensitivity is not high enough to detect the aggregates. In order to overcome these issues, CE-SDS with LIF detection was used. The rhuFab in one of the five formultions was derivatized with 5-TAMRA to yield the highest sensitivity and analyzed in CE-SDS/LIF (Fig. 10a). It is clear that the amount of free light/heavy chain is not being over estimated; whereas the aggregates were detected as well. The profile of the same sample after stressed at 40 °C for 6 months is shown in Fig. 10b. The amount of both the fragment and aggregrates increased as results of thermal stress. The example shown here clearly demonstrated the potential of CE-SDS/LIF with pre-column derivatization in terms of its capability of detecting protein stability and therefore can be used as a highly sensitive tool in the formulation screening/development of protein therapeutics.

Precolumn labeling with 5-TAMRA.SE combined with CE-SDS analysis is shown to be a highly sensitive approach for monitoring consistency of manufacture of therapeutic recombinant monoclonal antibodies based on their molecular weight dis-

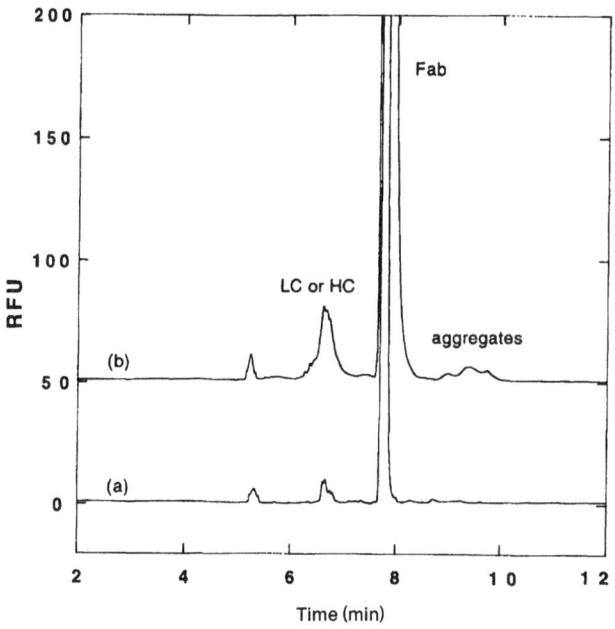

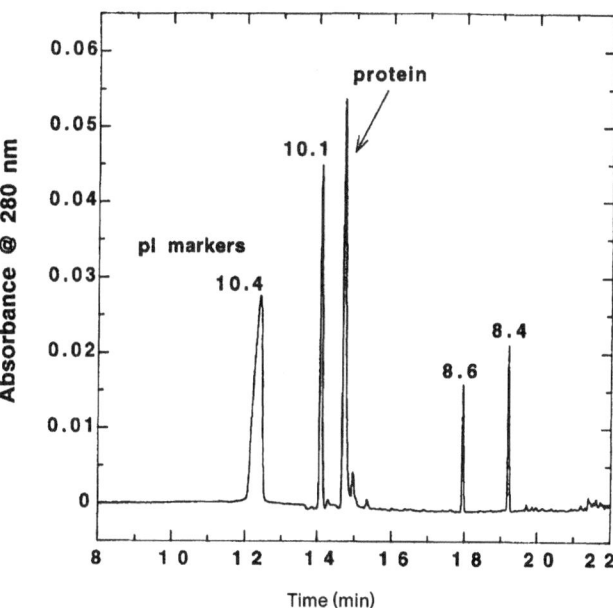

Figure 10. Electropherograms of rhuFab samples obtained by CE-SDS/LIF. Samples: (a) control; (b) after holding at 40 °C for 6 months. Capillary: fused silica capillary (50 μm × 27 cm); buffer: CE-SDS run buffer (Bio-Rad); voltage: 15 kV, reversed polarity; capillary temperature, 20 °C; injection: electrokinetic at 10 kV for 15 sec; detection, LIF with a 3.5 mW argon ion laser, 488 nm excitation/560 ± 20 nm emission.

Figure 11. Capillary isoelectric focusing profiles of a therapeutic protein with p*I* markers. The p*I* markers used here are the BioMark synthetic p*I* markers.

tribution. A demonstrated comparability in separation and sensitivity to SDS-PAGE visualized with silver stain, and the added advantages of speed, automation, and on-line detection make CE-SDS an acceptable replacement for the traditional slab gel approach in the quality control of recombinant therapeutic proteins. In addition, rapid analysis and robustness obtained with this technique vs SDS-PAGE also greatly reduce the methods development/validation time, and allow for faster routine product testing, resulting in increased productivity. Currently, several CE-SDS /LIF assays are being used as part of a control system for the release of rhuMab pharmaceuticals at Genentech in two select but critical applications: (a) monitoring the consistency of manufacture based on size distribution, and (b) detecting protein impurities, whether they be host cell proteins or product-related fragments and aggregates. Such assays have undergone complete validation in accordance with the guidelines of the International Committee on Harmonization (ICH) [23] and have been approved by the FDA and the European regulatory agencies for use as part of the lot release control system of marketed recombinant protein therapeutics.

9.3 Capillary Isoelectric Focusing (cIEF)

Capillary isoelectric focusing (cIEF) is another capillary version of the more conventional slab gel technique, in this case, the slab gel isoelectric focusing. The separation principle remains exactly the same with the exception that everything happens inside a capillary instead of the gel: a stable pH gradient is formed using carrier ampholytes and the proteins are focused to their individual isoelectric point. Since the detection in most commerical CE instruments is at a fixed point towards the outlet of the capillary, the focused protein zones have to be mobilized to pass through the detection window. This mobilization process can be achieved by means of hydrolytical, chemical mobilization and it often is the cause for loss of resolution and reduced reproducibility. Recently, an imaged capillary isoelectric focusing has been introduced as a "one-step" solution wherein the mobilization step is eliminated [24]. In this case, the isoelectric focusing still occurs inside the capillary, a CCD camera is used for detection along the whole capillary. More information about the cIEF technique and its various applications can be found in several review articles [25–27]. Among various applications, the two most commonly used applications for protein thera-

peutic analysis are p*I* determination and charge heterogeneity monitoring.

9.3.1 Determination of Isoelectric Point

The isoelectric point (p*I*) of an unknown protein is typically determined based on isoelectric point markers. Traditionally, pure proteins of known p*I* values are used as markers. However, the protein heterogeneity introduced in the initial production and/or after long-term storage of these markers hampers the accurate determination of the p*I* values. Over the past decade, low molecular weight amphorteric compounds have been introduced and commerialized as alternative p*I* markers to replace the traditionally used protein markers. These compounds have well defined p*I* values ranging from 5.3 to 10.4, strong UV absorbance at 280 nm and excellent stability under normal storage conditions. An example of the cIEF profile of a protein with unknown p*I* and four p*I* markers is illustrated in Fig. 11. The p*I* values of the markers are then plotted against their migration times in Fig. 12. It is seen from the figure, the pH gradient generated here is linear with a regression coefficient of 0.99. Based on the migration time of the unknown protein, its apparent p*I* value was easily calculated to be 9.8, which is near to the theoretical value of 9.9.

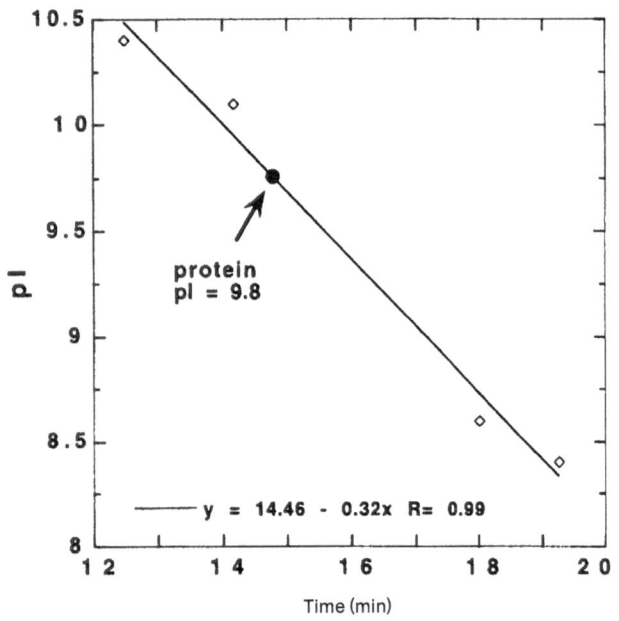

Figure 12. Determination of the p*I* value of a therapeutic protein by cIEF.

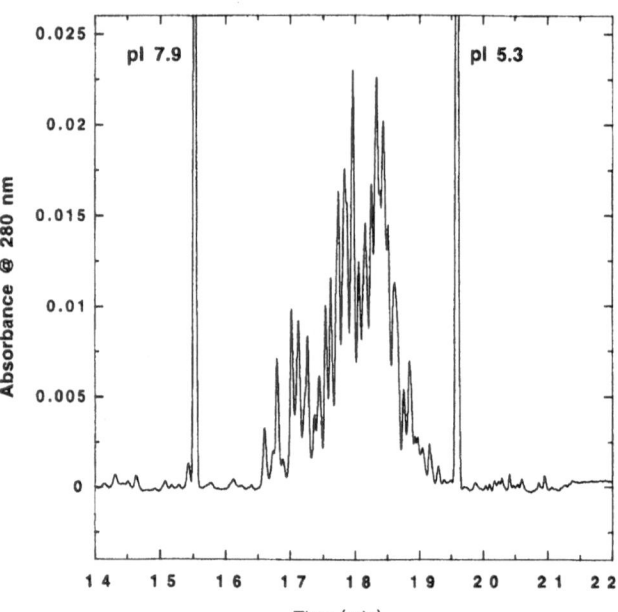

Figure 13. Capillary isoelectric focusing (cIEF) of a glycoprotein. Capillary: Bio-Rad Bio-CAP XL capillary (50 μm × 24 cm); ampholyte: 80% cIEF Bio-Lyte Ampholyte 3-10 (2% solution with 0.5% TEMED, 0.2% HPMC), 20% Bio-Lyte Ampholyte 3-10 (2%, diluted from 40% with H₂O); anolyte, 20 mM phosphoric acid; catholyte, 40 mM sodium hydroxide; focusing: 15 kV (625 V/cm) for 5 mins; mobilization: 20 kV (833 V/cm) for 25 mins with zwitterion (cathodic mobilizer from Bio-Rad); capillary temperature, 25 °C.

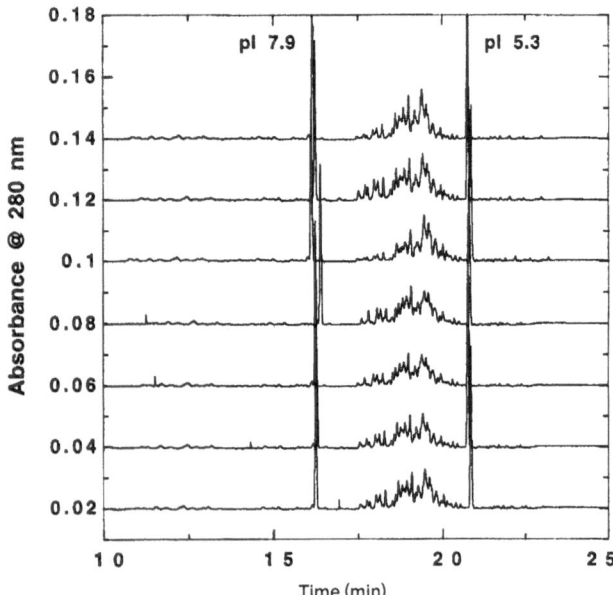

Figure 14. CIEF profiles of seven lots of glycoprotein manufactured at one manufacturing site. Experimental conditions are the same as listed in Fig. 13.

9.3.2 Monitoring Protein Charge Heterogeneity

The isoelectric point is also a fine indicator for any charge heterogeneity existing within a protein. Recent report [28] on the application and validation of a cIEF assay to monitor the charge heterogeneity of a recombinant monoclonal antibody is one example of its utility in our industry. The high resolution power of cIEF can be further examplified in its use for highly glycosylated, i.e., sialyated and phosphorylated glycoproteins, where microheterogeneity of the carbohydrates has significant contribution to the overall pro-tein charge variants. An example of the cIEF profile of a highly glycosylated protein is presented in Fig. 13. This glycoprotein is known to have cleavage at the C-terminal and deamidated species. In addition, post-translational modification on the protein resulted in approximately 3 moles of sialic acid per mole of protein. It is seen from the figure that the protein is resolved into about 28 major peaks representing the different charge variants. Due to the heterogenous nature of the protein, two p*I* markers are used to bracket the protein variants to best represent the p*I* span of almost two pH units. It is clear from the resolution power demonstrated here that cIEF profile can serve as a fingerprint in order to monitor the manufacturing consistency of a given product. Seven lots of the glycoprotein produced at one manufacturing facility were analyzed by cIEF as shown in Fig.14. It is seen from the figure that overall the protein is consistent from lot-to-lot in terms of the span of the p*I* range of the protein charged variants. There is, however, a slight change in the distribution among the variants resulting from the minute differences in the post-translational modification from lot-to-lot that is typically observed in the production of such glyco-proteins.

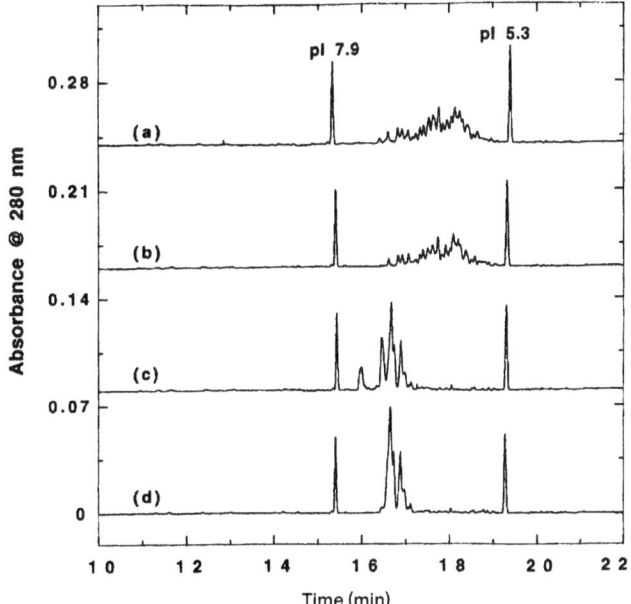

Figure 15. CIEF profiles of the glycoprotein obtained before or after different enzymatic digestions: (a) intact; (b) carboxypeptidase B; (c) neuraminidase; (d) carboxypeptidase B and neuraminidase. Experimental conditions are the same as listed in Fig. 13.

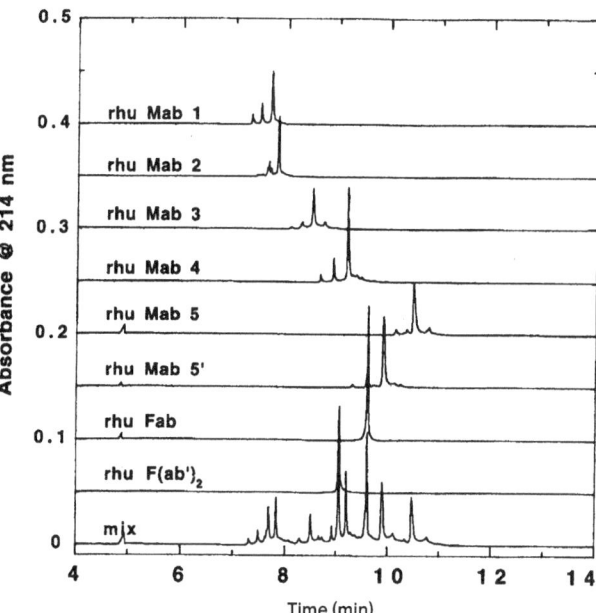

Figure 16. Capillary zone electrophoresis of eight recombinant human monoclonal antibodies/antibody fragments. Capillary: BioCAP XL coated capillary (50 μm × 47 cm); 45 mM ε-amino-*n*-caproic acid/acetic acid, pH 4.5, 0.1% HPMC; voltage: 30 kV, normal polarity; capillary temperature, 20 °C; detection, UV at 214 nm.

The cIEF profile of the intact protein is informative in terms of the overall protein charge distribution, which is a combination of the heterogeneity resulting from the protein backbone and the carbohydrates during the post-translational modification. In order to shed some light on the origin of the charge heterogeneity, carboxypeptidase-B (CpB) and neuraminidase are employed to remove the C-terminal lysine residue and the sialic acids, respectively. The cIEF profiles of the protein before and after enzymatic digestions are shown in Fig. 15. It is clear from the figure that the majority of the charge heterogeneity is attributed to the sialic acid distributions (Fig. 15c) as the number of peaks reduces dramatically down to a cluster of 4 from 28. In addition, the protein becomes more basic after removing the sialic acids with an estimated p*I* increase of one pH unit. The enzyme carboxypeptidase-B, on the other hand, removes the charge heterogeneity arising from the C-terminal lysine processing. As seen from Fig. 15b, the earliest eluting peaks which are the most basic variants disappeared as they became more acidic after losing the C-terminal lysine residue. This is even more clearly demonstrated in Fig. 15d, where the protein was treated with both enzymes. Clearly, the two basic peaks shown in Fig. 15c disappeared after the CpB digestion, which are believed to be the protein with the lysine at one or both C-termini.

9.4 Capillary Zone Electrophoresis (CZE)

Capillary zone electrophoresis (CZE) is the simplest and yet most versatile mode of all CE separations. Whereas the proteins are separated solely based on their p*I* differences in cIEF, the underlying separation mechanism in CZE is based on the electrophoretic mobility which is a function of both protein charge and its hydrodynamic size at a given environment [29]. Electrophoretic mobility can be manipulated by changing environment or separation conditions by means of changing pH, buffer additives, etc. Several reports have been published using CZE for the analysis of recombinant human growth hormone [30], recombinant human deoxyribonuclease [31] and recombinant human IGF-1 [32].

Recently, recombinant human monoclonal antibody or antibody fragment has gain increasing importance as therapeutic agents. Consequently, there is an increasing demand for generic assays with high resolution and yet high throughput to support product development. A generic CZE assay was developed for the analysis of rhuMabs to monitor protein charge heterogeneity in cell culture and recovery process development. The profiles of 6 rhuMabs, 1 rhuFab and 1 rhu F(ab')₂ are illustrated in Fig. 16. The proteins used here are fairly similar in terms of their p*I*

values ranging from 7.5 to 9.3. Moreover, two of the antibodies Mab 5 and 5' are the first and second generation of a therapeutic agent differing in only 5 AA residues. Yet, all the proteins are separated with baseline resolution under one generic condition in less than 12 minutes (Fig. 16, bottom trace). With a specific migration time that is unique to each product, these proteins can be easily identified based on the main peak's migration time. This can be used as a generic and rapid method to distinguish recombinant proteins from each other as a simple and yet specific final product and labeling identity test.

It is clearly seen from the profiles of these six full-length monoclonal antibodies that the various protein charge variants are also resolved. An expanded view of the profile for rhuMab 4 is shown in Fig. 17a. Further analysis with carboxypeptidase B digestion confirmed the three major species resolved here are related to the C-terminal lysine processing. With both lysine residues at the C-termini, the protein has the highest positive charges at the pH used here and therefore the highest mobility. Hence, the protein variants with two, one and zero C-terminal lysine residues elute in the order of increasing migration times. This example demonstrated the resolution power of capillary zone electrophoresis and its scope as a complementary technique to conventional ion-exchange chromatography with an added

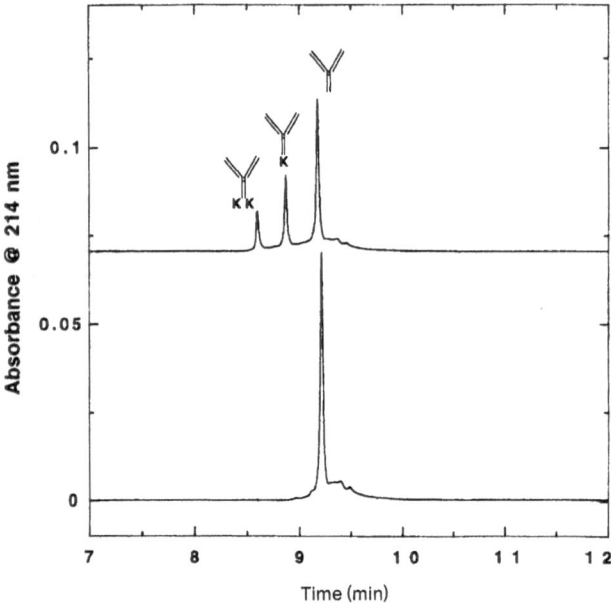

Figure 17. Capillary zone electrophoresis of a recombinant human monoclonal antibody: (a) intact (b) after digestion with Carboxypeptidase-B. Experimental conditions are the same as listed in Fig. 16.

advantage of rapid analysis time suitable for a high throughput process development environment.

9.5 Carbohydrate Analysis

There has been an increasing interest in the fundamental understanding of the biological roles of the carbohydrate moieties of recombinant glycoproteins in recent years. In contrast to the protein amino-acid backbone which is genetically coded with high fidelity, the cDNA sequence can only predict the position of potential glycosylation sites with little information about the actual glycosylation patterns. In fact, the biosynthesis of glycans on any glycoprotein is largely influenced by the cell line in which it is produced, the conformation of the protein and cell culture conditions. Even under a defined set of expression and culture conditions, further processing of the carbohydrate chains by enzymes present at limited quantities result in microheterogeneity of structurally related oligosaccharides at any of the glycosylation sites on a given glycoprotein. Such alterations in protein glycosylation with variable site occupancy or changes in oligosaccharide structure often result in variations in the biological activity of glycoproteins. The importance of the glycan structures in the therapeutic use of recombinant proteins has been well documented [33]. These biological functions and clearance are dependent not on-

ly on the presence or absence of certain preserved N-linked carbohydrate chains, but also the specific structure of any attached carbohydrates or even the specific terminal monosaccharide structure. Clearly, in the biotechnological manufacturing of therapeutic proteins, the assessment of oligosaccharides microheterogeneity and its batch-to-batch consistency are of utmost importance [34, 35]. Consequently, there is a growing demand for developing new and sensitive high performance analytical techniques to bring about the structural characterization and routine analysis of the closely related glycans derived from glycoproteins. While nuclear magnetic resonance and more recently mass spectrometry [36] are indispensable tools for the structural elucidation of carbohydrates, the routine profiling and quantitative analysis of glycans is accomplished mostly by chromatographic [37–41] and planar electrophoretic techniques [42]. Two of the most widely used techniques are high pH anion exchange chromatography with amperometric detection [38, 43] and fluorophore assisted carbohydrate electrophoresis [44, 45]. More recently, capillary electrophoresis has emerged as a powerful tool in carbohydrate analysis with enhanced separation efficiencies and shorter analysis time over the traditional methods. The fast implementation of CE in carbohydrate analysis was largely due to the fact that CE is simply the instrumental version of planar electrophoresis employing detection sys-

tems adapted from HPLC. As such, many of the electrolyte systems that were originally tested in traditional electrophoresis as well as precolumn derivatization schemes which afforded sensitive detection of carbohydrates separated by HPLC have been readily adapted for carbohydrate analysis by CE. Further progress in CE of carbohydrates was facilitated by the introduction of on-column laser-induced fluorescence (LIF) detection the most sensitive detection method employed to date. This also triggered the development of several tagging agents with superior fluorogenic properties. Among them, 8-aminopyrene-1,3,6-trisulfonic acid (APTS) [46, 47] has demonstrated numerous advantages over other commonly used reagents in terms of its detection sensitivity and speed of analysis. With the 488 nm argon-ion laser, the limit of detection for APTS-derivatized sugars was estimated to be 0.4 nM [47]. Therefore, it allows analysis of even minute carbohydrate content in therapeutic biomolecules of interest.

9.5.1 N-Linked Glycan Distribution

The glycans illustrated here are asialo- N-linked complex biantennary structures with a core fucose, typical of a monoclonal antibody produced in CHO cells. A schematic illustration of the glycans observed on a rhuMab is presented in Fig. 18. As seen from the figure, all glycans share the same fucosylated branched core structure but vary in their terminal galactose occupancy: (a) a degalactosylated glycan (G0), composed of eight sugar units with no terminal galactose; (b) a partially galactosylated glycan (G1), containing nine sugar units with one terminal galactose and hence two positional isomers; and (c) a fully galactosylated glycan (G2), containing ten sugar units having galactose on both termini. In order to monitor the glycan distribution in the production, a method was developed using a highly specific endoglycosidase, PNGase F, to remove the glycans enzymatically from the protein. The released glycans are subsequently derivatized with APTS at the reducing terminus. Without further purification, the derivatization mixture containing the excess reagents and the APTS-glycan adducts are simply diluted with water and analyzed by CE with on-column laser-induced fluorescence detection (LIF). A typical electropherogram for the analysis of the N-linked glycan distribution of this protein is shown in Fig. 19a and the section containing the glycans of interest is expanded in Fig. 19b. The multiple peaks observed between 1.5 and 2.5 minutes, also found in the APTS blank, are the excess APTS reagent and its impurity [47]. The APTS-glycan adducts and the internal standard (IS), migrate much later between 3.0 and 4.5 minutes. Since the APTS-glycans carry about the same net charges, the separation is based on the differences in their apparent hydrodynamic sizes. Therefore, the glycans migrate in the order of increasing size (G0 < G1 < G2) and are baseline resolved from each other. In addition, the two G1 positional isomers are also separated with baseline resolution with the first eluting peak as the G1(α1,6) [48]. The underlying mechanism for the separation of the two G1 positional isomers with the same molecular weight is based on the difference in their apparent hydrodynamic sizes resulting from the asymmetrical nature of the fucosylated branched core structure. The various aspects of assay development and validation were reported in detail [48]. Another example of this methodology is illustrated in Fig. 20. The glycans found on this monoclonal antibody contain high mannose structures (high mannose 5 and 6) in addtion to the complex type shown in Fig. 19. Using the separation buffer containing methylcellulose, the high mannose as well as the complex types of oligosaccharides were resolved under 5 minutes.

As carbohydrate plays a key role in many functions of a protein, it is clear that there is a need for new and improved analytical methodologies. The method illustrated here provides a simple, sensitive, rapid way of analyzing carbohydrates on various recombinant monoclonal antibodies or glycoproteins in general. In addition, the method was validated according to ICH guidelines [23] and was found to be accurate, precise and suitably robust for studies on N-linked carbohydrates and is currently being used for bulk product testing.

9.5.2 Monosaccharides

Another area of great interest is to explore further our understanding of the biosynthetic pathway of the oligosaccharides building blocks, i. e., the individual monosaccharide to provide some insight to the

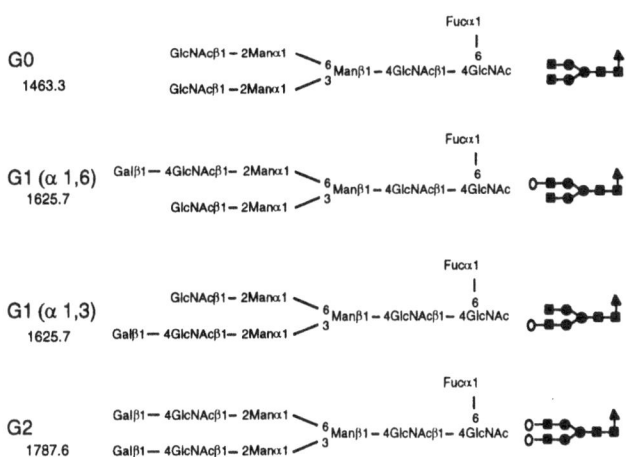
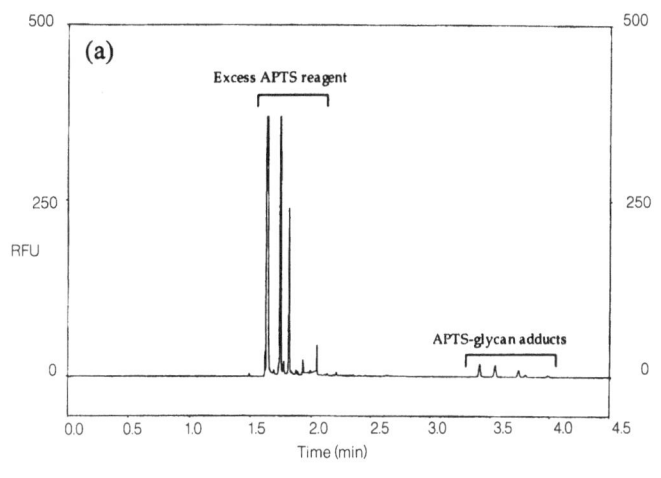
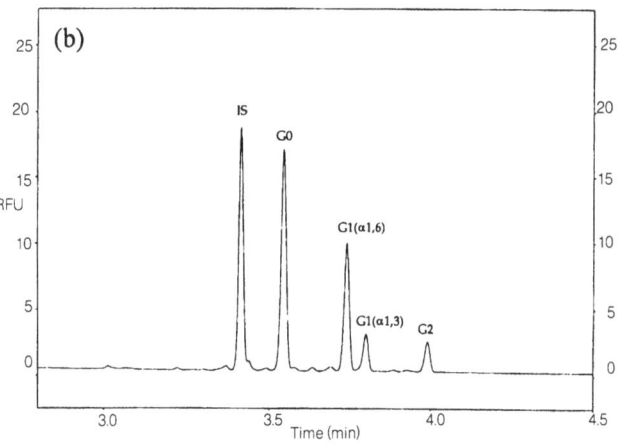

Figure 18. Structure of the *N*-linked glycans observed on a recombinant human monoclonal antibody. (From [48]. With permission.)

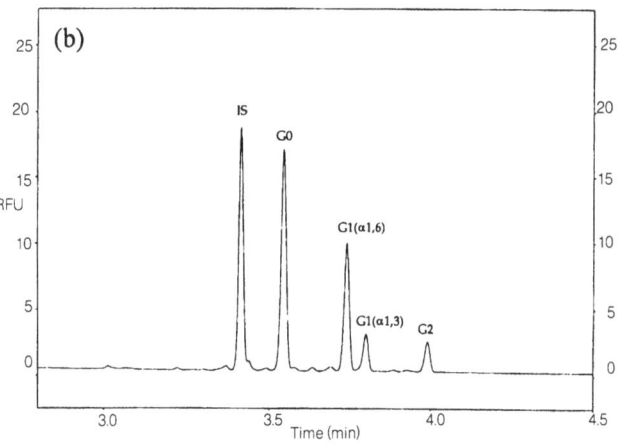

Figure 19. Typical electropherogram obtained (**a**) showing the glycan distribution; (**b**) an expanded scale view showing the glycans of interest. Capillary: Beckman N-CHO coated capillary (50 μm × 27 cm); carbohydrate gel separation buffer (Beckman Coulter); voltage: 20 kV, reversed polarity; capillary temperature, 20 °C; detection, LIF with a 3.5 mW argon ion laser, 488 nm excitation/520 ± 20 nm emission. (From [48]. With permission.)

understanding of the post-translational modification occurs in the cell culture. Sialic acids or α-*N*-acetylneuaminic acid (NANA) is often the terminal sugar for *N*-linked complex glycan structures. It is known to have an impact on the protein stability, activity as well as clearence of various glycoproteins [49]. As the intracellulor precursor for sialic acid synthesis is *N*-acetylmannosamine (ManNAc) [50–52], there is an interest to gain further understanding of the incorporation of sialic acid into glycoproteins produced in CHO cell lines by monitoring the ManNAc level

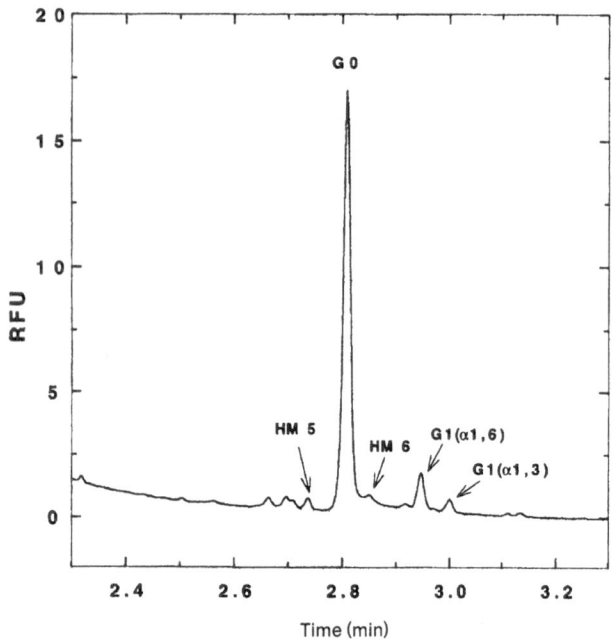

Figure 20. Profiles of *N*-linked oligosaccharides observed on a rhuMab with both complex and high mannose type structures. Capillary: Beckman N-CHO coated capillary (50 μm × 27 cm); 40 mM ε-amino-*n*-caproic acid/acetic acid, pH 4.5, 0.2% HPMC; voltage: 30 kV, reversed polarity; capillary temperature, 20 °C; detection, LIF with a 3.5 mW argon-ion laser, 488 nm excitation / 520 ± 20 nm emission.

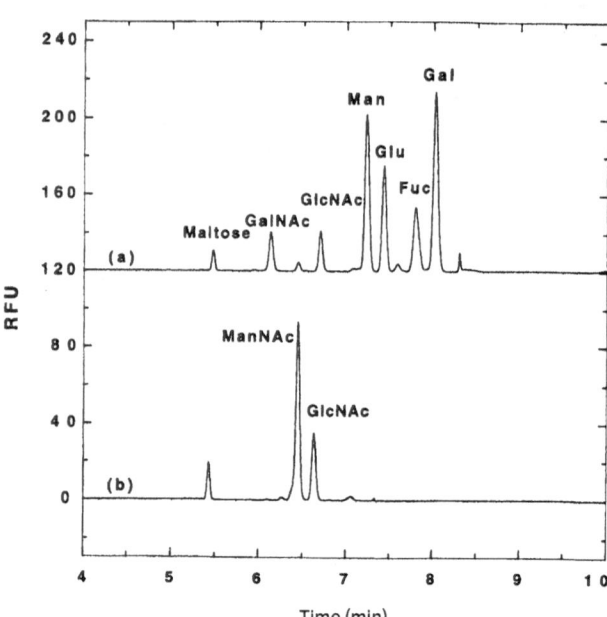

Figure 21. CE analysis of monosaccharides with borate complexation (a) typically found in CHO produced proteins; (b) ManNAc and GlcNAc. Capillary: fused silica capillary (50 μm × 27 cm); 100 mM sodium borate, pH 9.5; voltage: 10 kV, normal polarity; capillary temperature, 25 °C; detection, LIF with a 3.5 mW argon-ion laser, 488 nm excitation/520 ± 20 nm emission.

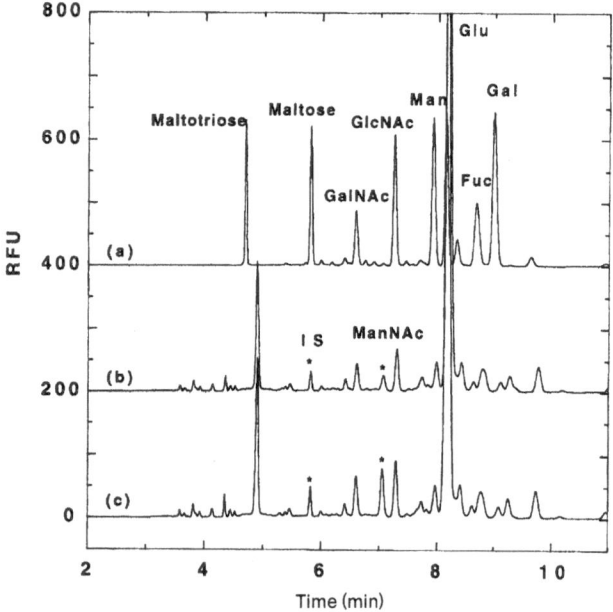

Figure 22. Electropherograms obtained for ManNAc spiked samples for the purpose of establishing the calibration curve: (**a**) standards; (**b**) control sample as the calibration blank; (**c**) 2 nmoles of ManNAc spike. Maltose (1 nmole) was spiked into the samples as the internal standard. Capillary: fused silica capillary (50 μm × 27 cm); 100 mM sodium borate, pH 9.5; voltage: 10 kV, normal polarity; capillary temperature, 25 °C; detection, LIF with a 3.5 mW argon-ion laser, 488 nm excitation/520 ± 20 nm emission.

in the cell culture fluid during fermentation. Since other monosaccharides, i. e., glucose, *N*-acetylglucosamine (GlcNAc), are expected to be present in the harvested cell culture fluids, a highly sensitive and specific assay is essential to the study. The separation of monosaccharides has been reported [36, 53, 54] by either LC or CE based on the differential complexation of monosaccharides with borate ions. In order to achieve the sensitivity level desired for the study, a CE assay with APTS labeling followed by laser induced fluorescence detection was developed.

Monosaccharides commonly found in eukaryotic glycoproteins were labeled with APTS and separated in borate buffer by CE as shown in Fig. 21a; whereas APTS labeled ManNAc standard was separated from all the six monosaccharides with baseline resolution as demonstrated in Fig. 21b. In order to quantitate the amount of ManNAc in terms of moles of ManNAc per grams of protein produced, a calibration curve of ManNAc was established. It has been shown previously in the literature [48] that it is crucial to keep the sample matrixes the same in the sample and the calibration standard in order to achieve similar labeling yields. Therefore, ManNAc standards were spiked into the control sample at different concentrations from 80 pmoles to 10 nmoles. In addition, an internal standard, maltose at a fixed concentration was added to each sample prior to the labeling reaction to account for any potential variation in the labeling and sample introduction. The electropherograms of the control sample and the 2 nmole ManNAc standard are shown in Fig. 22. As seen from the control sample profile, in addition to ManNAc, there are significant amounts of GalNAc, GlcNAc as well as glucose present in the HCCF. Hence, it should be pointed out the importance of having an assay that can resolve all these monosaccharides for this study. The relative corrected peak area (ManNAc/Maltose) of the control samples was subtracted from the ManNAc standards and a calibration curve was then established as illustrated in Fig. 23. It is seen from the figure that the calibration curve is indeed linear in the range from 80 pmole to 10 nmole with a linear regression coefficient better than 0.99. The fermentation time course samples were analyzed in the same fashion. An example of the electropherograms obtained at fermentation Day 1 and Day 3 are illustrated in Fig. 24. Subsequently, the ManNAc level as a function of the cell culture fermentation time course from Day 1 to Day 10 was determined and shown in Fig. 25. The synthesis of ManNAc can be then correlated to fermentation events, such as addition of certain nutrients. The methodology presented here is relatively simple and generic for other monosaccharides of interest. Yet, it can be applied for very sensitive detection of a specific monosaccharide in a rather complex mixture.

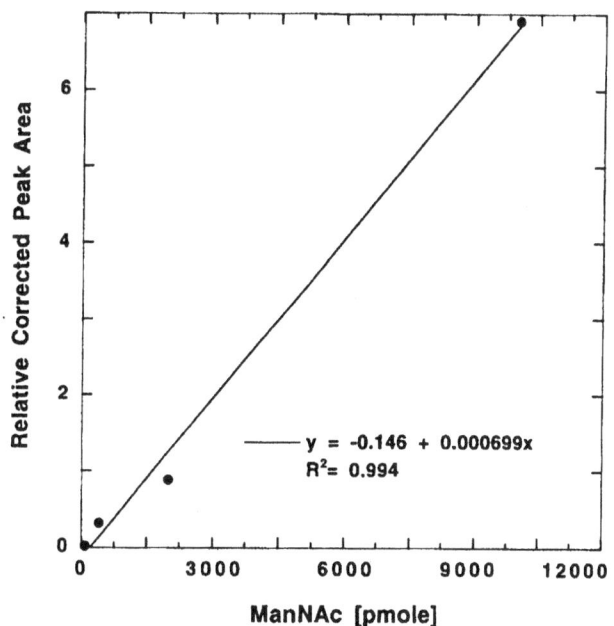

Figure 23. The relative corrected peak area of ManNAc to maltose is plotted as a function of the amount of spiked ManNAc to establish the calibration curve.

9.6 Methods

9.6.1 CE-SDS with UV Detection

Protein samples are typically diluted to at least $1\,mg\,mL^{-1}$ with the CE-SDS sample buffer (1% SDS in 100 mM Tris HCl, pH 9.2). For reduction, 5% (v/v) of β-mercaptoethanol is generally added to the sample. SDS-protein complexation is usually carried out at 60 °C for 10 to 20 minutes. The analyses are performed on a BioFocus capillary electrophoresis system (Bio-Rad). The samples are either hydrodynamically or electrokinetically injected into a fused silica capillary (50 μm ×

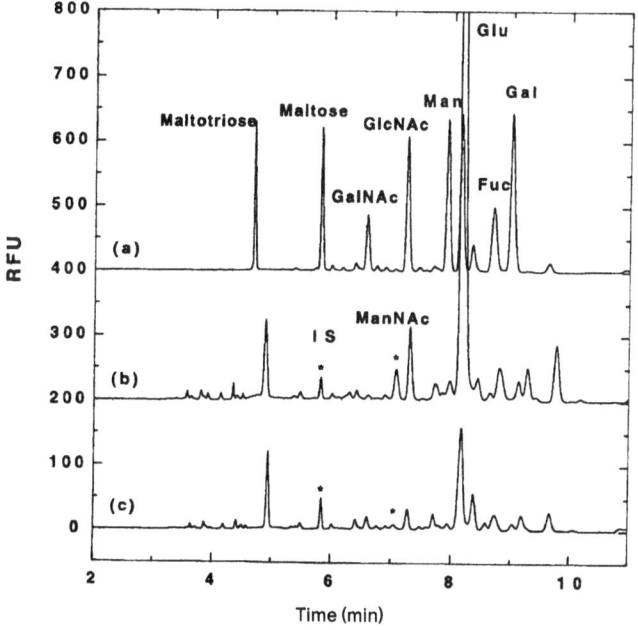

Figure 24. Electropherograms of monosaccharide distribution: (a) standards; (b) at Day 1 of the fermentation; (c) at Day 3 of the fermentation. Maltose (1 nmole) was spiked into the fermentation samples as the interanl standard for the analysis. Capillary: fused silica capillary (50 μm × 27 cm); 100 mM sodium borate, pH 9.5; voltage: 10 kV, normal polarity; capillary temperature, 25 °C; detection, LIF with a 3.5 mW argon-ion laser, 488 nm excitation / 520 ± 20 nm emission.

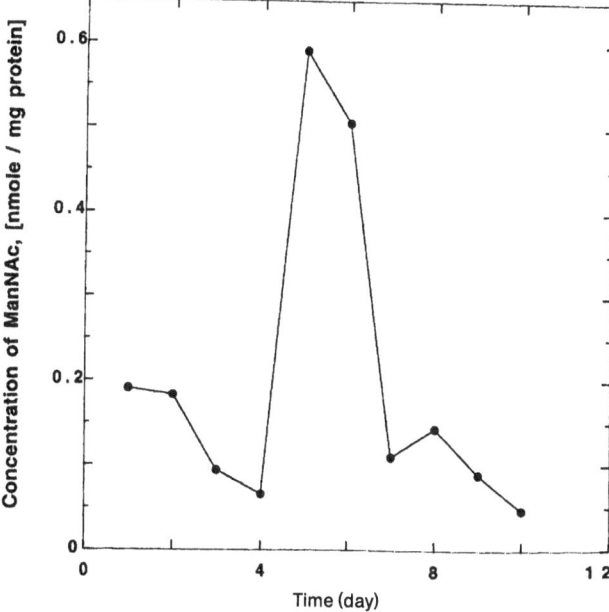

Figure 25. The concentration of ManNAc is plotted as a function of the fermentation time course.

24 cm) with capillary temperature set at 20 °C. The electrophoresis is carried out at a constant voltage of 15 kV in the reversedpolarity mode. The sieving medium consisted of a viscous cellulose solution at pH 8.3 and 0.5% SDS. The separation is then monitored oncolumn with UV detection at 220 nm.

9.6.2 CE-SDS with LIF Detection

Protein samples are buffer exchanged into 0.1 M sodium bicarbonate, pH 8.3. 5-TAMRA.SE dissolved in dry DMSO is then added to the protein solution and incubated at 30 °C for one to two hours. The dye-to-protein ratio typically ranges from 3 to 20 moles of dye to 1 mole of protein. After incubation, the excess dye is removed using NAP-5 columns. The SDS-protein complexation is carried out in the same way as previously described. The CE analyses are carried out in the same system as UV detection with the exception of using a laser-induced-fluoresence detection with an excitation wavelength of 488 nm from a 3 mW Argon-ion laser and the emission monitored at 560 nm (DF 40 nm).

9.6.3 Capillary Isoelectric Focusing – Protein's p*I* determination

Pharmalyte 8–10.5 (Pharmacia) was diluted to a final concentration of 2 wt%, 0.2% (v/v) hydroxypropylmethylcellulose (HPMC), and 0.5% (v/v) TEMED. The ampholyte used in the capillary isoelectric focusing consists of 80 % of the diluted pharmalyte 8–10.5, 20% of the cIEF Bio-Lyte 3–10 (Bio-Rad) with 0.1% of p*I* marker 8.6 (v/v). Samples were diluted to 1 mg mL^{-1} with the formulation buffer and then mixed with the cIEF ampholyte (10/90, v/v). The analyses were performed on a BioFocus 3000 capillary electrophoresis system (Bio-Rad). A BioCAP™ XL (Bio-Rad) coated capillary (50 μm ID × 24 cm) was used and the capillary temperature was held at 25 °C. Sample was injected by pressure at 100 psi for 40 seconds. The catholyte was 40 mM NaOH and the anolyte was 20 mM phosphoric acid. The focusing was carried out at a constant voltage of 15 kV for 6 minutes at the normal polarity. Mobilization was achieved using the Cathodic Mobilizer (Bio-Rad) at 20 kV for 20 minutes. The separation was monitored at 280 nm.

9.6.4 Capillary Isoelectric Focusing – Monitor Protein Charge Heterogeneity

Bio-Lyte 3–10 (Bio-Rad) was diluted to a final concentration of 2 wt%. The ampholyte used in the capillary isoelectric focusing consists of 20% of the diluted Bio-Lyte 3–10, 20% of the cIEF Bio-Lyte 3–10 (Bio-Rad), which contains 0.2% (v/v) hydroxypropylmethylcellulose, and 0.5% (v/v) TEMED, and 0.1% of p*I* marker 5.3 and 7.9 (v/v). All protein samples were diluted to 1 mg mL^{-1} with either the respective protein formulation buffer or water. All protein samples were mixed with ampholyte at the ratio of 2:1:1 (ampholyte:protein:water, v/v) prior to analysis. The analyses were performed on a BioFocus 3000 capillary electrophoresis system (Bio-Rad). A BioCAP™ XL (Bio-Rad) coated capillary (50 μm ID × 24 cm) was used and the capillary temperature was held at 25 °C. Sample was injected by pressure at 100 psi for 40 seconds. The catholyte was 40 mM NaOH and the anolyte was 20 mM phosphoric acid. The focusing was carried out at a constant voltage of 15 kV for 5 minutes at the normal polarity. Mobilization was achieved using the Cathodic Mobilizer (Bio-Rad) at 20 kV for 25 minutes. The separation was monitored at 280 nm.

9.6.5 Capillary Zone Electrophoresis

All protein samples were diluted to 0.1 to 1 mg mL^{-1} with water. The analyses were performed on a P/ACE 5000 or MDQ system (Beckman Coulter). A commercially available neutral, hydrophilic coated capillary (Beckman or Bio-Rad), with dimensions of 50 μm ID and a length from the inlet to detector of 40 cm, is used in all the cases shown here. The capillary temperature is held at 20 °C. Sample was injected by pressure at 0.5 psi for 10 sec (or 5 psi · sec). The separation buffer is 45 mM ε-amino-*n*-caproic acid with acetic acid, pH 4.5, 0.1% HPMC. The separation was monitored at 214 nm.

9.6.6 Carbohydrate Analysis – N-linked Glycan Distribution

PNGase F digestion. Typically antibody in the formulation buffer is buffer exchanged into the PNGase F digestion buf-

fer (20 mM sodium phosphate, pH 7.5, containing 50 mM EDTA, and 0.02% (w/v) sodium azide using a Microcon-30 concentrator. Five units of PNGase F are then added to the sample (about 2 nmoles of rhuMab) and incubated for approximately 15 hours at 37 °C. The deglycosylated protein is heated at 95 °C for 5 minutes and then precipitated by centrifugation at 10,000 × g for 10 minutes. The supernatant containing the oligosaccharides is dried in a centrifugal vacuum evaporator to a translucent pellet.

APTS Derivatization. The pellet is reconstituted in 19.1 mM solution of APTS in 15% acetic acid followed by the addition of sodium cyanoborohydride solution. The labeling solution is kept at 55 °C for 2 hours and then diluted approximately 25-fold with water prior to CE analysis.

Capillary Electrophoresis. A P/ACE 5000 CE system (Beckman) equipped with a 3-mW argon-ion laser with an excitation wavelength of 488 nm and an emission bandpass filter of 520 ± 20 nm is used in the study. Coated capillaries with reduced electroosmotic flow, the eCAP™ N-CHO capillaries from Beckman (Fullerton, CA, USA) or the BioCAP™ LPA and BioCAP™ XL capillaries from BioRad (Hercules, CA, USA) can be used. All capillaries are of 50-μm i. d. and 27/20-cm length with a capillary temperature of 20 °C. The samples are introduced hydrodynamically for 8 seconds at 0.5 psi. The separation is performed at a constant electric field of 740 V cm^{-1}. Between the runs, the capillary is rinsed with the buffer at 20 psi for 1 minute.

9.6.7 Carbohydrate Analysis – Monosaccharides

APTS Derivatization. The protein samples with addition of the internal standard, maltose, or additional ManNAc standards for the calibration curve are dried to completeness in a SpeedVac. The pellet is reconstituted in 190 mM solution of APTS in 1 M citric acid followed by the addition of 2 M aqueous solution of sodium cyanoborohydride. The labeling solution is kept at 60 °C for overnight and then diluted approximately 50-fold with water prior to centrifugation for 5 minutes at 10000 × g. The supernatant is recovery and subject to CE analysis.

Capillary Electrophoresis. Exactly the same instrument setup is used for the monosaccharides analysis. The capillaries used here are fused silica of 50-μm i.d. and 27/20-cm length. The capillary temperature is set at 20 °C. The samples are introduced hydrodynamically for 4 seconds at 0.5 psi. The separation buffer used is 100 mM sodium borate, pH 9.5 and the separation is performed in the normal polarity at a constant electric field of 370 V cm^{-1}. Between the runs, the capillary is rinsed with at least 5 column volumes of 0.1 N sodium hydroxide, water, the borate buffer.

9.7 Acknowledgements

The authors gratefully acknowledge Eleanor Canova-Davis, Tony Chen and Andy Jones for valuable discussions and support over the years. The authors would also like to acknowledge Glenn Hunt, Brandon Tomlinson, Rob Fahrner, Sandra Sandall, Xanthe Lam, Nina Le, Namita Nayak, Rod Keck, Wendy Lau, Long Truong, Lori Schalk, Glen Teshima, and Tom Warner for their contributions.

9.8 References

[1] Weber, K.; Osborn, M. *J. Biol. Chem* **1969**, *244*, 4406–4412.

[2] Reynolds, J.A.; Tanford, C. *J. Biol. Chem.* **1970**, *245*, 5161–5165.

[3] See, Y.P.; Jackowski, G. In *Protein Structure: A Practical Approach*; Creighton, T.E., Ed.; IRL Press: Oxford, **1989**, pp 1–21.

[4] Shapiro, A.L.; Vinuela, E.; Maizel, J.V. *Biochem. Biophys. Res. Commun.* **1967**, *28*, 815–820.

[5] Chrambach, A.; Reisfeld, R.A.; Wyckoff, M.; Zaccari, J. *Anal. Biochem.* **1967**, *20*, 150–153.

[6] Oakley, B.R.; Kirsch, D.R.; Morris, N.R. *Anal. Biochem.* **1980**, *105*, 361–363.

[7] Merril, C.R.; Switzer, R.C.; Keuren, M.L.V. *Proc. Natl. Acad. Sci. U.S.A.* **1979**, *76*, 4335–4339.

[8] Hjerten, S. *J. Chromatogr.* **1983**, *270*, 1–6.

[9] Hjerten, S. In *Electrophoresis '93*; Hirai, H., Ed.; Walter de Gruyter: New York, **1984**, pp 71–79.

[10] Cohen, A.; Karger, B.L. *J. Chromatogr.* **1987**, *397*, 409–417.

[11] Karger, B.L.; Paulus, A.; Cohen, A.S. *Chromatographia* **1987**, *24*, 15–22.

[12] Zhu, M.; Hansen, D.L.; Burd, S.; Gannon, F. *J. Chromatogr.* **1989**, *480*, 311–319.

[13] Ganzler, K.; Greve, K.S.; Cohen, A.S.; Karger, B.L.; Guttman, A.; Cooke, N.C. *Anal. Chem.* **1992**, *64*, 2665–2671.

[14] Nakatami, M.; Shibukawa, A.; Nakagawa, T. *J. Chromatgr. A* **1993**, *672*, 213–218.

[15] Guttman, A.; Nolan, J.; Cooke, N. *J. Chromatogr.* **1993**, *632*, 171–175.

[16] Guttman, A. *Electrophoresis* **1996**, *17*, 1333–1341.

[17] Takagi, T. *Electrophoresis* **1997**, *18*, 2239–2242.

[18] Hunt, G.; Nashabeh, W. *Anal. Chem.* **1999**, *71*, 2390–2397.

[19] Rasmussen, J.R. *Current Opinion in Structural Biology* **1992**, *2*, 682–686.

[20] Varki, A. *Glycobiology* **1993**, *3*, 97–130.

[21] Brinkley, M. *Bioconjugate Chem.* **1992**, *3*, 2–13.

[22] Haugland, R.P. *Handbook of Fluorescent Probes and Research Chemicals*; Molecular Probes, Inc.: Eugene, OR, **1996**.

[23] International Conference on Harmonization: Guideline on the validation of analytical procedures: methodology. *Fed. Regist.* **1997**, *62 (96)*, 27464–7.

[24] Wu, J.; Li, S.-C.; Watson, A. *J. Chromatogr. A* **1998**, *817*, 163–171.

[25] Rodriguez-Diaz, R.; Wehr, T.; Zhu, M.; Levi, V. In *Handbook of Capillary Electrophoresis*; Landers, J., Ed.; CRC Press: Boca Raton, 1997, pp 101–188.

[26] Wehr, T.; Zhu, M.; Rodriguez-Diaz, R. In *Methods in Enzymology*; Karger, B.L., Hancock, W.S., Eds.; Academic Press: San Diego, 1996; Vol. 270, pp 358–374.

[27] Righetti, P.G.; Gelfi, C.; Chiari, M. In *Capillary Electrophoresis in Analytical Biotechnology*; Righetti, P.G., Ed.; CRC Press: Boca Raton, 1996, pp 509–539.

[28] Hunt, G.; Hotaling, T.; Chen, A.B. *J. Chrom. A* **1998**, *800*, 355–367.

[29] Kalman, F.; Ma, S.; Fox, R.; Horvath, C. *J. Chromatogr. A* **1995**, *705*, 135–154.

[30] McNerney, T.; Watson, S.; Sim, J.; Bridenbaugh, R.L. *J. Chromatogr. A* **1996**, 223–229.

[31] Felten, C.; Quan, C.; Chen, T.; Canova-Davis, E.; McNerney, T.; Goetzinger, W.K.; Karger, B. *J. Chromatogr. A* **1999**, *853*, 295–308.

[32] Nashabeh, W.; Greve, K.; Kirby, D.; Foret, F.; Karger, B.; Reifsnyder, D.; Builder, S. *Anal. Chem.* **1994**, *66*, 2148–2154.

[33] Parekh, R.B.; Dwek, R.A.; Sutton, B.J.; Fernandes, D.L.; Leung, A.; Stanworth, D.; Rademacher, T.W.; Mizuochi, T.; Taniguchi, T.; Matsuta, K.; Takeuchi, F.; Nagano, Y.; Miyanato, T.; Kotata, A. *Nature* (London) **1985**, *316*, 452–457.

[34] Olden, K.; Bernard, B.A.; White, S.L.; Parent, J.B. *J. Cell. Biochem.* **1982**, *18*, 313.

[35] Kobata, A. *Eur. J. Biochem.* **1992**, *209*, 483.

[36] El Rassi, Z.; Nashabeh, W. In *Carbohydrate Analysis: HPLC and CE*; ElRassi, Z., Ed.; Elsevier: New York, **1995**, pp 267–360.

[37] Honda, S.; Suzuki, S. *Anal. Biochem.* **1984**, *142*, 167–174.

[38] Hardy, M.R.; Townsend, R.R.; Lee, Y.C. *Methods Enzymol.* **1989**, *179*, 65–76.

[39] Tandai, M.; Endo, T.; Sasaki, S.; Masuho, Y.; Kochibe, N.; Kobata, A. *Archives of Biochemistry and Biophysics* **1991**, *291*, 339–348.

[40] Lipniunas, P.; Grönberg, G.; Krotkiewski, H.; Angel, A.-S.; Nilsson, B. *Archives of Biochemistry and Biophysics* **1993**, *300*, 335–345.

[41] El Rassi, Z. In *Journal of Chromatography Library*; Elsevier: New York, **1995**; Vol. 58.

[42] Morell, L.; Plotkin, L.; Leoni, J.; Fossa, C.A.; Margni, R.A. *Molecular Immunology* **1993**, *30*, 695–700.

[43] Spellman, M. *Anal. Chem.* **1990**, *62*, 1714–1722.

[44] Jackson, P. *Analytical Biochemistry* **1991**, *196*, 238–244.

[45] Starr, C.; Masada, R.; Hague, C.; Skop, E.; Klock, J. *J. Chromatogr. A* **1996**, *720*, 295–321.

[46] Chen, F.-T.A.; Evangelista, R.A. *Anal. Biochem.* **1995**, *230*, 273–280.

[47] Evangelista, R.A.; Liu, M.-S.; Chen, F.-T.A. *Anal. Chem.* **1995**, *67*, 2239–2245.

[48] Ma, S.; Nashabeh, W. *Anal. Chem.* **1999**, *71*, 5185–5192.

[49] Cumming, D.A. *Glycobiology* **1991**, *4*, 115–130.

[50] Gu, S.; Wang, D.I.C. *Biotechnology and Bioengineering* **1998**, *58*, 642–648.

[51] Ferrari, J.; Gunson, J.; Lofgren, J.; Krummen, L.; Warner, T.G. *Biotechnology and Bioengineering* **1998**, *60*, 589–595.

[52] Santell, L.; Ryll, T.; Etcheverry, T.; Santoris, M.; Dutina, G.; Wang, A.; Gunson, J.; Warner, T.G. *Biochemical and Biophysical Research Communications* **1999**, *258*, 132–137.

[53] Foster, A.B. *Adv. Carbohydr. chem.* **1957**, *12*, 81–116.

[54] El Rassi, Z. *Adv. Chromatogr.* **1994**, *34*, 177–250.

Sensitive and High Resolution CE/MS/MS for Protein Identification in Complex Mixtures

W. Tong[1] / J. R. Yates[2]

[1] EPIX Medical, Inc., Cambridge MA, USA
[2] Department of Molecular Biotechnology, University of Washington, Seattle, WA 98115, USA

10.1 Introduction

Capillary electrophoresis (CE) as a high-resolution separation technique emerged from the pioneering work of Hjertén [1]. It took more than a decade to develop the modern version of capillary zone electrophoresis (CZE) [2, 3]. The high surface-to-volume ratio of capillaries allows for efficient dissipation of Joule heat generated from the high electric field applied. As a result, separations are much faster and of higher resolution than those of slab gel electrophoresis. Separations are created by the electrophoretic mobility of the analyte in the presence of electroosmotic flow (EOF). The flow profile of analyte in CE is flat instead of parabolic as in a pressure driven system like HPLC. This endows CE with high separation efficiency (high number of theoretical plates). In addition, the low sample consumption associated with the narrow bore capillary in CE (a few nL) makes its absolute sensitivity extremely high. All these features make CE a very attractive separation tool for the analysis of proteins and peptides [4, 5].

Interfacing CE with mass spectrometry began with the coupling of CE to continuous-flow fast-atom bombardment (cf-FAB) sources [6–9]. These efforts were quickly supplanted by interfaces to electrospray ionization sources. ESI is a gentle ionization technique that generates intact molecular ions. The distribution of multiply charged ions generated by ESI allows high molecular weight species to be detected within the mass range of most mass spectrometers. Smith et al reported the first direct coupling of CE with ESI-MS [10, 11]. All commercially available CE/MS interfaces at present are based on ESI. Ten years after this pioneering work, mass spectrometry has developed to be one of the most important detectors for CE [12]. Besides its unsurpassed selectivity, the low flow rates with narrow-bore capillary size facilitate the formation of a shorter jet from which direct emission of droplets can be obtained [13]. Attomole-level detection limit has been reported for peptides [14, 15]. Highly sensitive techniques to analyze peptides and proteins will provide an important interface to biological discovery driven by genome projects.

Large-scale whole-genome sequencing of organisms is building an information database that will change the breadth and scale of biological experiments. Genome sequence information can be used to help sort out the function of the gene products or proteins. One method used to separate and quantitate complex mixtures of proteins is two-dimensional gel electrophoresis (2-DGE). Mass spectrometry has become an increasingly important tool to identify the separated proteins [16–18]. This has been made possible through the development of computer algorithms to identify proteins by correlating the molecular weights of collections of peptides derived from proteolytic digests or tandem mass spectra of individual peptides with sequences in the database [19–22]. Large-scale analysis of proteins has been termed Proteomics.

Among the challenges of Proteomics small quantity (sensitivity) and complexity (resolution) of the samples are probably the most prominent. Highly sensitive staining methods are used to visualize proteins separated by two-dimensional gel electrophoresis. For example, the sensitivity of silver staining is approximately 1–5 ng, which corresponds to 20–100 fmol for a 50 kD protein. Sensitive microscale separation/tandem mass spectrometry techniques have been developed to match the sensitivity of silver staining. Many cellular processes are performed and regulated by proteins acting in macromolecular complexes that are often composed of large numbers of unique proteins. Methods to identify complex mixtures of proteins without the need to purify each component to homogeneity not only improves the efficiency of protein identification, but should also increase the sensitivity of detection. CE/MS, with its ability to handle extremely small amounts of samples with high separation efficiencies, is well suited for sensitive identification of proteins in complex mixtures [23]. In this chapter, various approaches for enhancing the sensitivity and resolving power of CE/MS will be discussed.

10.2 Approaches to Improve the Sensitivity of CE/MS/MS for Protein Identification

10.2.1 Interface Design

Coupling CE to electrospray ionization requires a closed circuit to drive the electrospray process. Three major types of interfaces have been designed for coupling

0009-5893/00/02   90-10 $ 03.00/0

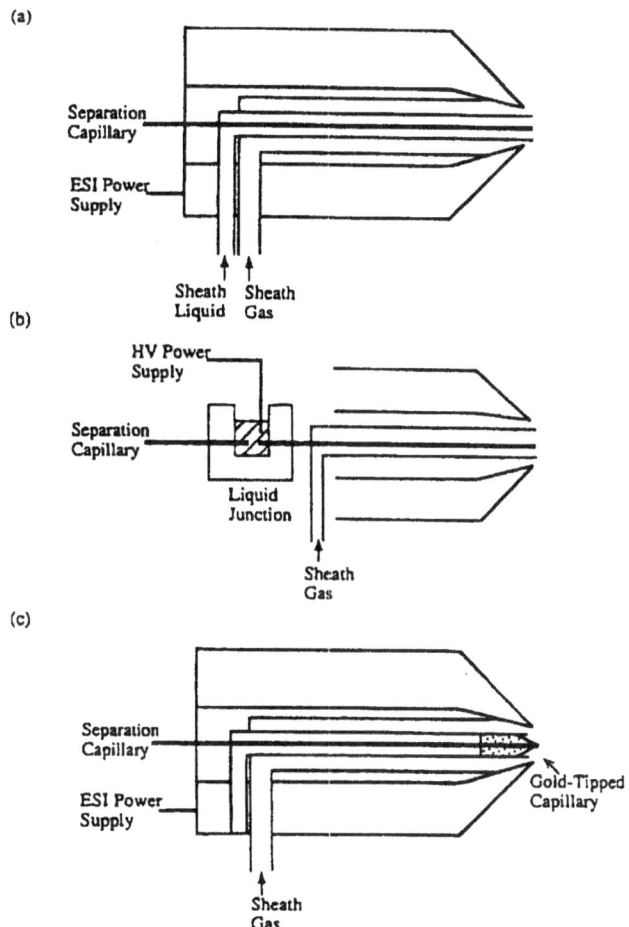

(a)

Separation
Capillary

ESI Power
Supply

Sheath   Sheath
Liquid   Gas

(b)

HV Power
Supply

Separation
Capillary

Liquid
Junction

Sheath
Gas

(c)

Separation
Capillary

ESI Power
Supply

Gold-Tipped
Capillary

Sheath
Gas

Figure 1. Schematic diagram of three major types of CE/ESI-MS interfaces. (a) Coaxial sheath flow interface; (b) Liquid junction interface; (c) Sheathless interface. (Reprinted with permission from *Handbook of Capillary Electrophoresis*, 2nd Ed., James P. Landers, Ed., 1997, pp 794, Copyright 1997 by CRC Press, Inc.)

of CE with MS (Figure 1). The coaxial sheath-flow interface (Fig. 1a), which is also the basis for most commercially available interfaces, was first reported by Smith, et al. [11]. Although it is the most commonly used and can be easily and reproducibly constructed, two major drawbacks limit its sensitivity for peptide analysis in CE/MS. First, the introduction of a higher flow rate sheath liquid dilutes the CE eluent at the mixing point. The increased volume of eluent decreases detection sensitivity and separation efficiency. Second, the sheath liquid will introduce additional background that can reduce the signal to noise. In addition, the formation of moving ionic-boundaries inside the capillary leads to variations in migration times, resolution and changes in migration orders [24].

The liquid-junction interface (Fig. 1b) reported by Henion and co-workers [25] establishes the electrical contact through a liquid reservoir in which analytical capillary and ESI sprayer are carefully butted

end-to end. This interface accommodates different flow rates, ranging from those of nano-ESI to those of pneumatically assisted ESI. Precise alignment of the analytical capillary and ESI sprayer is critical to minimize peak broadening and decrease of sensitivity due to the sample loss at the junction.

The original version of sheathless interface was reported by Olivares et al. [10], using a metal coating on the tip of the capillary to apply ESI voltage and close the CE circuit. In principle, the sheathless interface is the simplest for coupling CE to nano- or microspray because of the compatible flow rates of each technique. It has been shown that with small i. d. capillaries (5–10 µm) and careful selection of CE buffer, to minimize background signals, *S/N* values of the CE/micro-ESI-MS can be enhanced 25 to 50-fold over those for conventional ESI. Many efforts have been made since to improve the performance of the sheathless interface. Fang et al. [26] utilized a gold wire inserted into the outlet

of the CE capillary to provide electrical contact. Ramsey et al. [27, 28] reported using a silanized or chromium coated capillary tip to improve the stability of the gold coating for electrical contact. Protein identification at subfemtomole levels using gold-coated sprayer for CE/MS/MS has been reported [29].

The most recent versions of sheathless interfaces are actually based on the liquid junction. Smith and co-workers used a short length of microdialysis tubing, across which electrical contact can be made, to connect CE capillary to a short micro-ESI sprayer [30]. Again, the precision and ruggedness of the alignment of the CE capillary and ESI sprayer are critical to maintain the high sensitivity and separation efficiency of the sheathless interface, given the fact that the microdialysis tubing is flexible enough to cause misalignment. Thibault and co-workers [31] described two approaches to the production of the nanospray CE/MS sheathless interface: (1) one-piece column tapered at one end and gold-coated and (2) tapered gold-coated tips butted to the CE column (Figure 2). Several methods for connecting the butted tips to the CE column were investigated with the Supelco connector producing the same separation efficiency as the one-piece nanospray column but required less time in construction. In fact, these microsprayers are prepared from a length of capillary with the inner wall coated with suitable reagents and subsequently tapered to suitable dimensions (7–10 cm long). A large number of disposable microsprayers can thus be prepared together.

A more rugged liquid-junction sheathless interface was made by butting together CE capillary and non-gold-coated spray tip through a short piece of stainless steel tubing [32]. Since epoxy was used to seal the liquid junction, the microsprayer was not easily replaced. Tong et al. [23] reported a sheathless interface using a finger-tight PEEK micro-tee as the liquid junction. It is a modified version of previously reported sheathless interface for micro-LC [33]. A gold wire was inserted in the side channel of the micro-tee to provide electrical contact for ESI voltage. The microsprayer is easily replaced. With the PEEK-tee, the background associated with the stainless steel and epoxy at low pH could be minimized. Subfemtomole detection limits are obtained with this interface for protein identification using CE/MS/MS (Figure 3). To minimize the

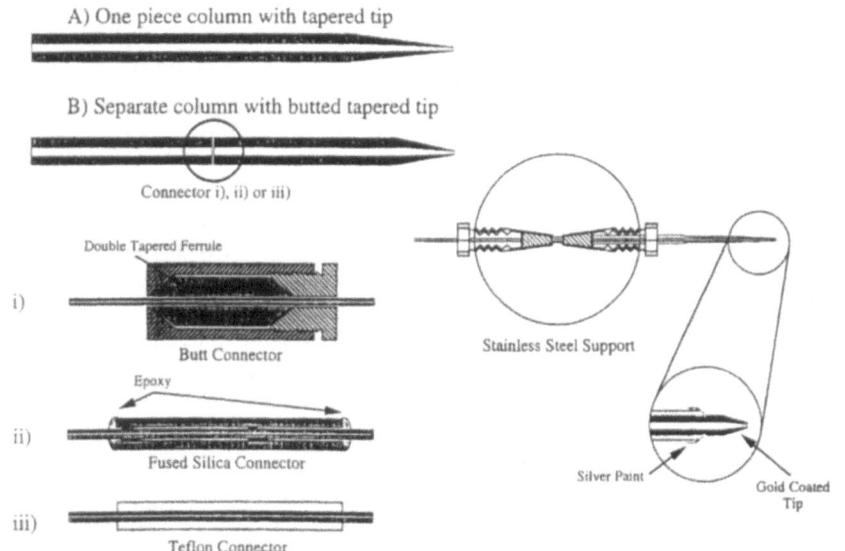

Figure 2. CE/ESI-MS nanospray interface arrangements. For the butted-tips, connection is made before the stainless steel support using connector **i)**, **ii)** or **iii)**. (Reprinted with permission from *Rapid Communications in Mass Spectrometry*, Vol. 11, pp 309, **1997**, John Wiley & Sons, Ltd., New York).

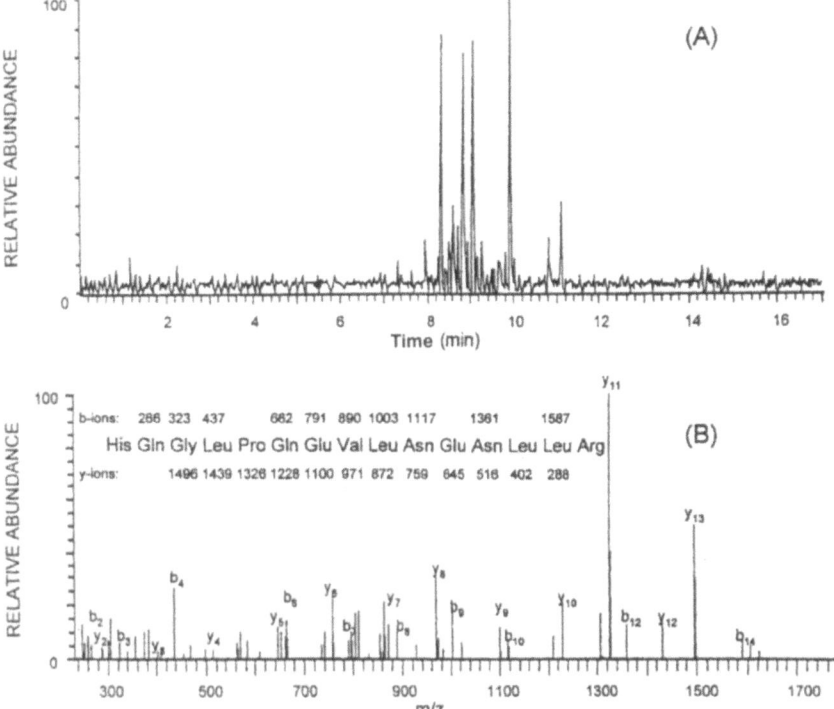

Figure 3. (A) CE/MS/MS base peak electropherogram of 600 amol of α-casein tryptic digest. (B) Collision-induced dissociation mass spectrum for $(M+2H)^{2+}$ ion of HQGLPQEVLNENLLR peptide at *m/z* 881. (Reprinted with permission from *Anal. Chem.* **1999**, *71*, pp 2274.)

dead volume and interruption of CE caused by bubble formation at the ESI electrode, Cao et al inserted a Pt wire into the CE capillary through a small hole cut near the terminus [34]. However, a small amount of methanol (5–10%) in the CE running buffer can virtually eliminate bubble formation during a normal CE run (30–40 min) in liquid junction sheathless interfaces [23, 31, 32]. Between runs, the capillary can be flushed with running buffer to purge accumulated bubbles in the system. Since the CE capillary and the sprayer are precisely butted together in liquid-junction sheathless interfaces [23, 31, 32], the dead volume is negligible and separation efficiency is not affected. Figure 4 shows a separation of a mixture of four

test peptides using this system [35]. In our experience, the liquid-junction sheathless interface is the most convenient and practical means to achieve highest sensitivity in CE/MS/MS for protein identification.

Tjaden and co-workers reported a novel micro-ESI interface for CE/MS that uses neither sheath liquid nor an electrode. [36]. The outlet of the CE capillary is tapered to serve as a microspray tip. Electrospray is created by the electrical field of the CE separation voltage and the inlet to the mass spectrometry serves as the counterelectrode. The electrical contact at the capillary outlet is established through air between the spray tip and the inlet of the mass spectrometer, which is at ground potential. A one order of magnitude increase in sensitivity was observed over a conventional sheath-flow interface.

10.2.2 Microfabricated Devices

Microfabricated devices are of interest because of the ability to integrate multiple functions or operations in a single device. The intent of "lab-on-the chip" devices is to increase the speed, throughput and sensitivity of analyses [37]. Small quantities of sample can be injected, diluted or preconcentrated, mixed, separated, and detected in a highly efficient and reproducible manner with these devices. Early efforts to couple microfabricated devices with electrospray ionization have focused on sample manipulation on the chip followed by off-chip infusion driven by pressure or electroosmosis [38–41]. Although promising results have been achieved with these approaches, the use of an external device to promote fluid delivery compromise some of the advantages of the microchip. The ability to electrokinetically manipulate small sample volumes, for example, may be lost or impaired. In addition, there is a large dead volume associated with the droplet formed at the ESI exit port in infusion driven chips [38–40], and this prevents the use of such a design for performing on-chip CE/MS. The use of a transfer capillary is a good means to overcome this weakness [41]. Most recent designs using flat-tip drilled hole and precisely butted spray tip can virtually eliminate the conical dead volume of the droplet [42]. Both sheathless ESI [43] and on-chip pneumatic nebulizing assisted ESI [43, 44] interfaces using microfabricated devices for CE/MS. Protein identification using chip-CE/MS/MS was demonstrated

[44]. Aebersold and co-workers demonstrated the application of automated multi-channel electroosmotic pumping [45] and nanoflow solvent gradient delivery microfabricated devices [46] for the identification of proteins separated by 1D and 2D gel electrophoresis using MS/MS. The most recent report by Ramsey and co-workers [47] demonstrated subattomole to low-attomole sensitivity within millisecond time frames by using microchip sheathless nanospray ESI sources with time-of-flight mass spectrometry detection. Previous reports could achieve similar sensitivity only with nanospray sources using relatively long ion accumulation/injection times (seconds) [48, 49].

10.2.3 On-line Preconcentration

CE offers excellent absolute detection limits because extremely small amounts of sample can be injected (nL). Its concentration detection limit, however, is relatively poor requiring sample concentrations in the range of $\geq 10^{-6}$ M of sample. Biological samples are usually very dilute with a relatively large volume (μL). Therefore, preconcentration (preferably on-line) is needed to make CE a useful technique for "real world" samples. Two types of online preconcentration, based on electrokinetic concentration and solid phase extraction (SPE), respectively, have been successfully used for CE/MS.

On-line transient isotachophoretic preconcentration is the most commonly used method of electrokinetic concentration. A sample is sandwiched between two buffer zones, the front zone is a leading electrolyte (LE), which has a faster mobility than the sample ions, and the rear zone consists of a terminating electrolyte (TE) with lower mobility than the sample ions. Karger and co-workers [50, 51] applied the technique for trace analysis of proteins and improved the concentration detection limit of full scan CE/MS by a factor of 100 to ~10^{-7} M.

Capillary isoelectric focusing (cIEF) is a high-resolution separation technique that separates proteins based on their isoelectric point. Because of the focusing effect, analyte is concentrated into narrow bands resulting in an increase in concentration by a factor of 50–100. On-line coupling of cIEF with MS with a coaxial sheath-liquid flow assisted interface could handle protein concentration on the level of 10^{-7} M [52]. cIEF/MS is better used for

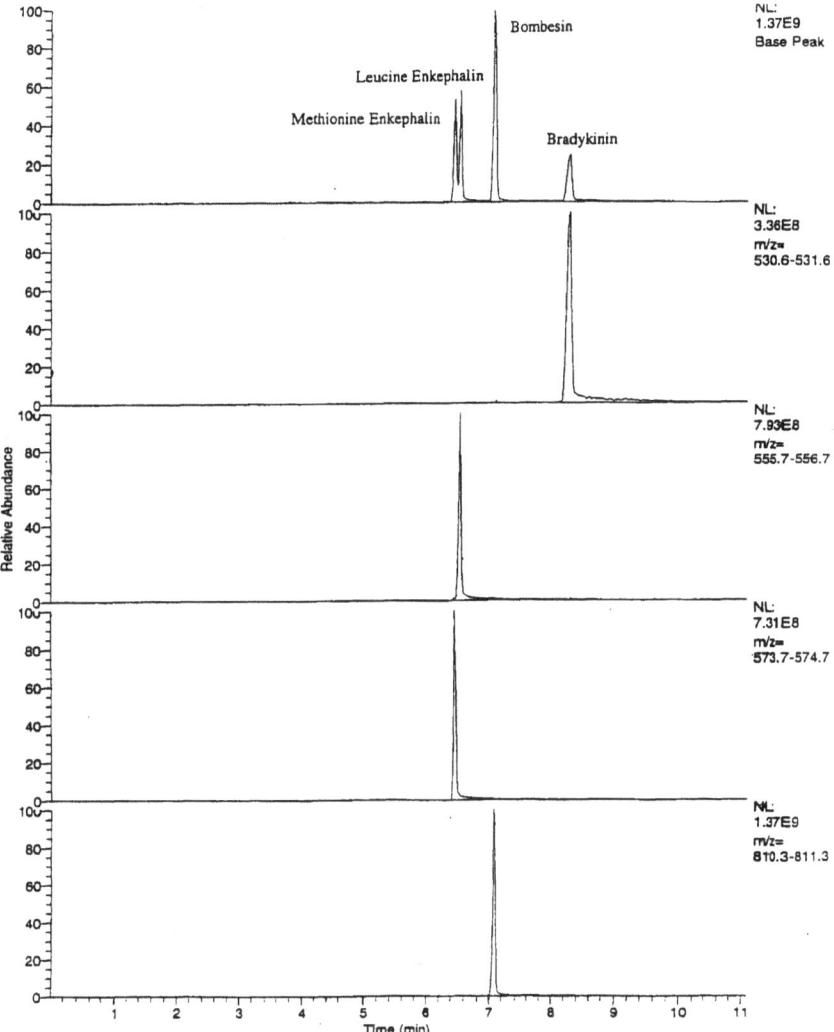

Figure 4. CE/MS base peak and extracted ion electropherograms of four-peptide mixture using a liquid-junction sheathless interface with precisely butted microspray tip and CE capillary in a finger-tight PEEK micro-tee. Amount of each peptide injected: methionine enkephalin: 240 fmol; leucine enkephalin: 240 fmol; bombesin: 85 fmol; bradykinin: 130 fmol.

proteins than for peptides because the ampholytes contribute to background in the low end of the mass range. To reduce the background a method for on-line desalting of proteins separated by cIEF/MS was also reported [53].

An effective method to preconcentrate a dilute mixture of peptides is on-line solid phase extraction (SPE). Two types of SPE have been introduced; packed chromatography beds [54–57] and hydrophobic membranes [58–61]. By using preconcentration on hydrophobic media, the sample is not only concentrated but salts, buffers, and other small-molecule contaminants are removed. Figure 5 shows the schematic of membrane preconcentration-CE-MS setup [61]. The packed bed device is similar except two pieces of polymeric frit material are used to contain the packing beads. A large volume (10's of μL) of solution can be passed through SPE material

concentrating the peptide mixture at the head of the CE column. After washing with CE running buffer, a plug of organic phase is injected to elute the concentrated sample. The column is then returned to the running buffer for normal CE separation and MS detection. Peak broadening is associated with the spatial distribution introduced during elution of the peptides off the solid phase. Peaks can be sharpened by using moving-boundary transient isotachophoresis to further focus the analyte zone [56]. Reports have shown the concentration detection limit can be improved at least 1000 fold to the low atto-mole μL^{-1} range [32]. Membrane-based devices are easier to construct than a device packed with chromatographic material. Beds packed with chromatographic material offer a wider variety of phases and hence selectivity since there are many different types of chromatographic phase

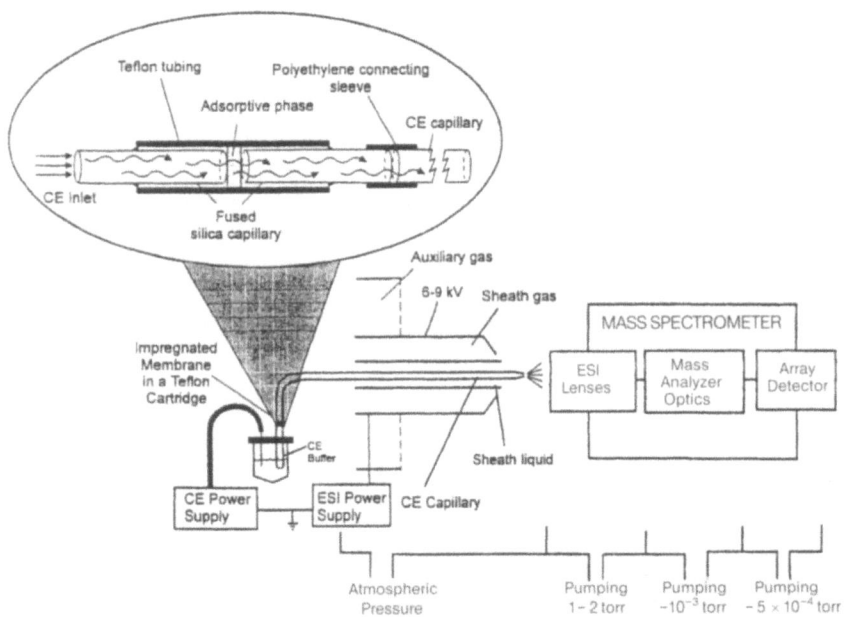

Figure 5. Schematic of membrane preconcentration/CE/ESI-MS. (Reprinted with permission of Elsevier Science from "Utility of membrane preconcentration-capillary electrophoresis-mass spectrometry in overcoming limited sample loading for analysis of biologically derived drug metabolites, peptides, and proteins", by Tomlinson, A.J., Benson, L.M., Jameson, S., Johnson, D.H., Naylor, S., *J. Am. Soc. Mass Spectr.*, **1997**, *8*, No. 1, pp 17).

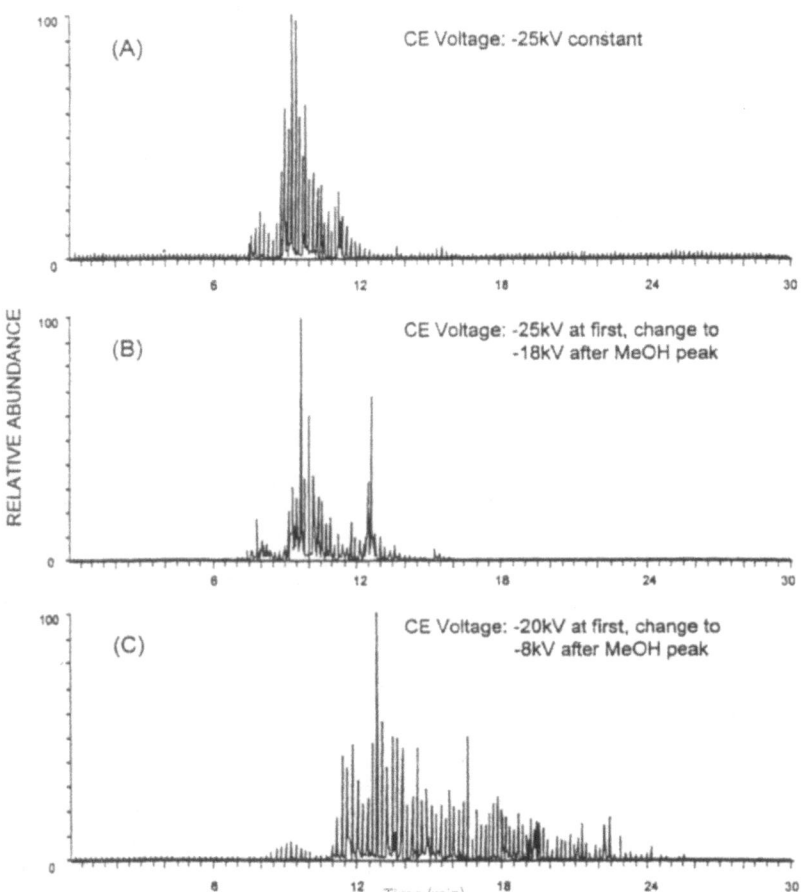

Figure 6. Micro-SPE/CE/MS/MS traces for the tryptic digest of yeast ribosomal complex at different separation voltages. (**A**) CE was run at −25 kV constantly after the elution step. (**B**) CE was run at −25 kV after the elution, and the voltage was reduced to −18 kV after methanol peak. (**C**) CE was run at −20 kV after elution, and the voltage was reduced to −8 kV. Elution were performed by injecting 80% methanol (0.5% acetic acid) at −25 kV for 45 s. (Reprinted with permission from *Anal. Chem.* **1999**, *71*, pp 2275.)

available. The SPE/CE/MS/MS method has been demonstrated for the identification of low abundance proteins isolated from 2-D gel [32, 54]. Characterization of major histocompatibility complex (MHC) class I peptides and clinically derived proteins using SPE/CE/MS/MS and SPE/CE/MS was also reported [61]. SPE/CE/MS/MS has also been used to identify sites of protein modification [62]. The fluorescently labeled acrylodan was determined to be covalently bound to lysine-27 of intestinal fatty acid binding protein by using membrane-preconcentration/CE/MS/MS.

10.3 Approaches to Improve the Resolution of CE/MS/MS for Protein Identification

10.3.1 Reduced Elution Speed CE, Lower EOF Coating and Dynamic Exclusion MS

CE is a high resolution and high-speed separation technique based on the difference of ion mobility in an electric field. Since ion mobilities of peptides are reasonably similar, they tend to have relatively close migration times, resulting in a small peak capacity. For complex mixtures, all components may not completely resolve even though peaks in CE are sharp. Because the duty cycle of the MS/MS for present mass spectrometers is on the order of seconds (1–3 seconds), the fast elution of peptides does not allow sufficient time to perform MS/MS on each ion. Smith and co-workers [63] reported a method they named "reduced elution speed CE" where the electrophoretic voltage is decreased prior to the elution of the first analyte peak. By decreasing the CE voltage, electrosmosis is decreased. Slowing down the elution of solutes allows more scans to be recorded during peak elution without a significant loss in ion intensity. This is particularly important for MS/MS methods for protein identification. Tong et al [23] applied this method for the identification of proteins in the yeast ribosome complex using micro-SPE/CE/MS/MS. Figure 6 shows the difference in constant voltage CE and reduced elution speed CE. For constant voltage CE, most peptides elute off the column over a 5 minutes window. Database searching using SEQUEST only identified 13 proteins with 15 unique peptide matches. By reducing the CE voltage from −25 to −8 kV after elution of methanol (EOF), peptides were spread over a 10-min time window and database search by SE-QUEST identified 39 proteins out of ~75 with 57 unique peptide matches.

For CE/MS of peptides and proteins, the inner wall of the capillary is usually coated with a positively charged film and separations are performed under acidic conditions. Proteins and peptides carry the same charge as the capillary wall and the peak broadening and tailing from electrostatic interactions with the capillary wall are minimized. Commonly used capillary coatings (such as (3-aminopropyl)-trimethoxysilane (APS)) for electrospray ionization of peptides [64, 65] produce relatively large electroosmotic flow (EOF), creating relatively short separation times that do not allow sufficient time to fully resolve analytes [31, 66]. Dynamic coating with cationic polymers such as Microcoat [67] and polybrene [68, 69] were reported to produce better resolutions than APS because they have the same effect to minimize the adsorption to the capillary wall but produce lower EOF. Lubman and co-workers used polybrene coated capillary for rapid analysis of hemoglobin variants with sufficient separation to resolve the large number of products from the whole protein so that prior separation of hemoglobin α and β chains is not required. One drawback for dynamic coatings like polybrene is the requirement for regenerating the inner surface between runs where high concentrations of acidic buffers (>0.1 M formic acid) are used. In addition, this kind of non-permanent coating can slowly peel off during CE runs and create higher background and thus lower the detection limit in MS. Covalently immobilized cationic coatings, such as [(acryloylamino)-propyl]trimethylammonium chloride (BCQ) and [3-(methacryloylamino)pro-pyl]trimethylammonium chloride (MAP-TAC), are thus used to solve this problem [31, 70].

Another strategy to coordinate the speed of CE and the duty cycle of MS/MS is to use data-dependent modulation [71]. Using an algorithm written in Instrument Control Language (ICL) which modulate CE voltage in a data-dependent manner, any ion which was detected in MS with an intensity exceeding a preset threshold value triggers the high-voltage power supply to reduce the flow of the analytes (park the peak) and instructed the MS to switch to the MS/MS mode. As an example, the MS/MS analysis of a minor phosphory-lated peptide from a proteolytic digest of

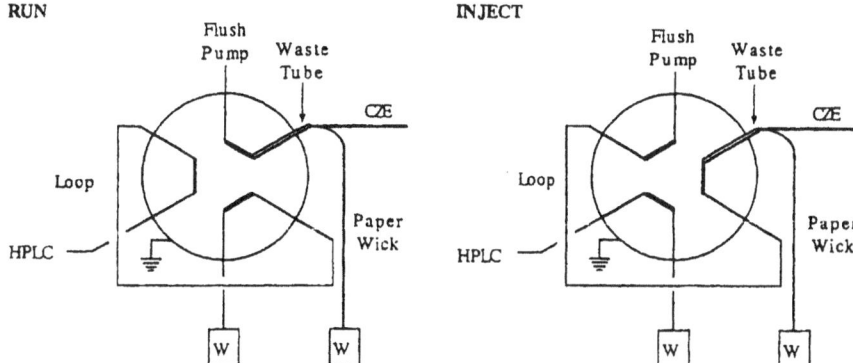

Figure 7. Schematic of a 6-port valve-loop interface for 2D microbore LC-CE. (Reprinted with permission from *Handbook of Capillary Electrophoresis*, 2nd Ed., James P. Landers, Ed., CRC Press, Inc. **1997**, pp 773).

endothelial nitric oxide synthase (eNOS) was demonstrated [71].

10.3.2 Multidimensional Separation Strategies

Direct analysis of large protein complexes (DALPC) enables comprehensive identification of individual proteins in macromolecular complexes without first purifying each protein component to homogeneity. Although proteins in simple mixtures can be identified by LC/MS/MS [72], complex mixtures will still overwhelm the resolution capability of any one-dimensional separation system. Multidimensional LC and tandem mass spectrometry has been successfully used for this application [73].

In fact, LC/CE/MS/MS should be ideal for this type of applications. CE/MS has proven to have better absolute sensitivity than LC/MS while the first dimension LC could also be used as a preconcentration step to make up the weakness of low sample loading for CE. Generally for 2D separations, the two separation systems should be orthogonal (i. e., with distinctly different mechanisms) and the sample transfer between the two dimensions should be efficient (without or little dilution, dispersion or recombination) [74]. In a comprehensive 2D separation, discrete fractions taken from the entire elution range of the first separation dimension are further separated/detected by the second dimension/MS. This generally requires a much faster second dimension in order to not miss any fractions from the first dimension. CE can be done in the order of seconds with shorter capillaries and higher voltage and well suited for this purpose.

Jorgenson pioneered the use of two-dimensional LC/CE [75]. Since 1990, they have published a series of papers showing the coupling of reversed phase LC (RPLC) [76–80], size-exclusion chromatography (SEC) [81] as well as SEC-RPLC [82]. Yamamoto et al also reported the coupling of gel permeation chromatography [83] and HPLC [84] with CE for the analysis of complex protein mixtures.

Microbore LC and CE are easily coupled through a 6-port electrically-actuated valve equipped with an external sample loop [76, 77] (Figure 7). In the run mode, the effluent from the LC column is collected in the 10-μL loop. When the valve is in the inject mode, the flush flow is directed to the loop and the collected components are electromigrated onto the CE capillary. The valve is grounded to serve as the CE anode. The CE capillary is connected directly to the valve through a waste tube. The waste tube is equipped with a paper wick to allow excess effluent to flow cleanly to waste.

To couple micro-LC and CE, Jorgenson developed a transverse flow-gating interface (FGI) [79] (Figure 8). Two 3-in. diameter, 0.5-in. thick stainless steel plates are separated by a Teflon gasket to allow liquid to flow between the plates. In a blocking position (Fig. 8A), transverse flush flow carries the micro-LC effluent to waste without entering the CE capillary. When the valve is actuated to the transfer position (Fig. 8B), the transverse flush flow is stopped and the micro-LC effluent fills the narrow gap of FGI. Electrokinetic injection is then performed by application of –1 kV to –5 kV. CE is initiated after the valve is returned to the blocking position and the transverse flow resumes. Injections onto the CE capillary are overlapped

A. Blocking Position

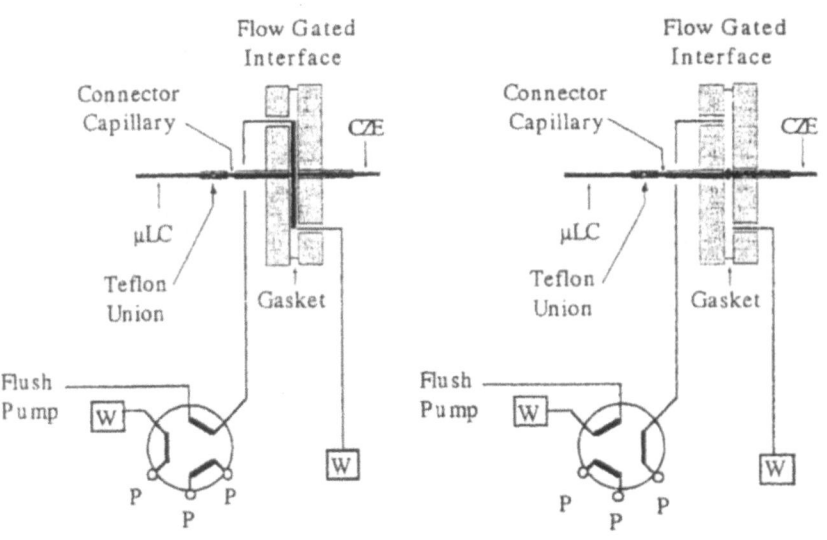

B. Transfer Position

Figure 8. Schematic of the transverse flow-gating interface for 2D micro-LC-CE. (Reprinted with permission from *Handbook of Capillary Electrophoresis*, 2nd Ed., James P. Landers, Ed., CRC Press, Inc. **1997**, pp 778.)

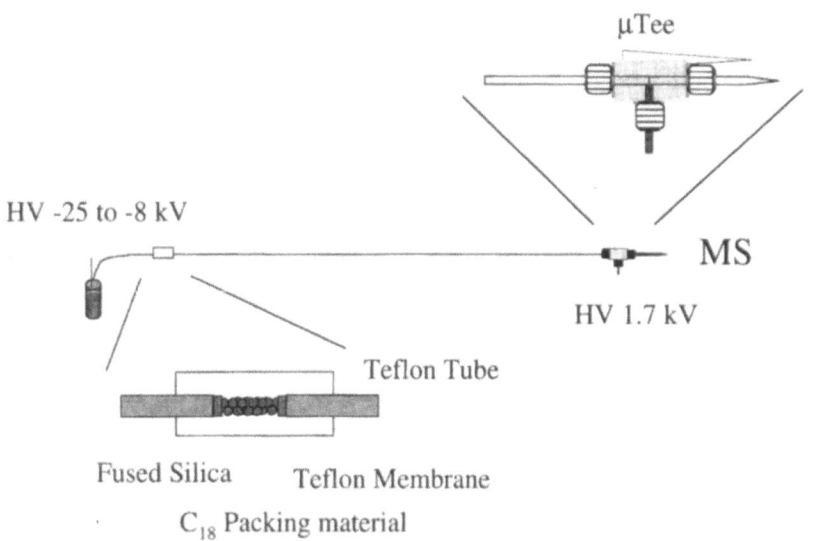

Figure 9. Schematic of micro-SPE/CE/ESI-MS interface for multistep elution/CE/MS/MS. (Reprinted with permission from *Anal. Chem.* **1999**, *71*, pp 2272.)

to use a majority of the 2D separation space and to sample the LC effluent more frequently. While most of these reports used protein tryptic digests and optical detection to test the separation power of the 2D system, on-line RPLC-CZE-MS system was also reported for peptide mixtures [85].

We reported a simple method of performing multidimensional microscale separation/tandem mass spectrometry for the identification of proteins in complexes [23]. It is an integration of preconcentration, multidimensional separation, reduced speed CE and sheathless ESI/MS/MS detection. The schematic setup is

shown in Figure 9. It consists of micro-SPE/CE coupling to a liquid-metal sheathless ESI interface. The method simply utilizes the micro-SPE as the first separation and CE as the second separation dimension. The resolution of micro-SPE is achieved through multistep elution by step gradients of organic solvent. Loading of sample onto the micro-SPE can be performed in-line with the mass spectrometer or off-line using a pressurized bomb at ~20 psi, followed by a cleanup by flushing with a small volume (~5 µL) of CE running buffer (0.5% acetic acid). A stepwise gradient of methanol (in 0.5% acetic acid) (usually from 10–80%) was used to elute

the peptides by injecting a plug of methanol. After injection of each gradient step, a CE/MS/MS run was performed. To minimize repeated acquisition of MS/MS on the same peptide ion, the dynamic exclusion function of the LCQ was enabled. Reduced elution speed CE was also employed to allow enough time to perform MS/MS on more peptide ions. Figure 10 shows the micro-SPE/CE/MS/MS traces for 10, 25, 40, 60, and 80% methanol step elution during a total of an 11-step elution run (10, 20, 25, 30, 35, 40, 45, 50, 60, 70, and 80% methanol in 0.5% acetic acid, successively). A total of 250 ng of the proteolytically digested yeast ribosomal complex was injected onto the SPE cartridge. The CE/MS/MS traces were different for each step elution, indicating a certain degree of separation occurred during the elution step gradient. For the 75-protein complex from yeast ribosome, 80–90% of the proteins in the complex was identified by searching the database using the MS/MS data with SEQUEST. Only 40–50% of the yeast ribosomal complex proteins could be identified with a single-step elution. As shown in Table I, the number of unique peptides and the total number of peptides scored are significantly increased for multistep elution SPE/CE/MS/MS compared to the single-step elution. In addition to its high sensitivity, the automation of the process could increase the speed of the overall analysis relative to multidimensional LC methods [73]. Also, more selective control over the CE running voltage can be obtained on some mass spectrometers to control the CE voltage in a data-dependent manner [71]. The concept of micro-SPE could be extended by the use of other types of separation materials in combination with reversed-phase packing material to effect greater selectivity for multidimensional CE separations [86].

10.4 Other Approaches

Capillary electrochromatography (CEC) is a hybrid method of capillary HPLC and CE [87]. In CEC, solvent is delivered by electroosmotic flow instead of the hydraulic flow that occurs in HPLC. The separation efficiency increases because of the plug-flow profile of the EOF and the ability to use smaller particles than those used in HPLC to pack the column. Despite these advantages, the separation of proteins and peptides has not been extensively explored. When coupling CEC with MS, the proteins and peptides exhibit better detection limits using positive ESI mode at low pH, which will greatly reduce the number of ionized silanol groups that are essential in generating EOF. Also, the residual ionized silanol groups will interact with the positively charged peptides resulting in peak broadening. For this reason, APS coating and pressure assisted flow have to be used. Lubman and co-workers [88, 89] have reported the coupling of open-tubular CEC and pressurized CEC with MS for peptide mixture analysis. Recently, porous monolithic columns has been used for CEC. Horváth and co-workers [90] described monolith columns in situ formed inside the capillary by polymerizing a mixture of vinylbenzyl chloride and divinylbenzene in the presence of a porogen. The chloromethyl groups of the polymeric surface were reacted with N,N-dimethyl-dodecylamine to form a positively charged hydrophobic stationary phase. This is compatible with the low pH separation and positive ESI detection of MS for proteins and peptides. If the smaller stationary phase particles ($\leq 1.5\,\mu m$) are packed or the monolith columns are prepared in narrower bore capillary ($\leq 30\,\mu m$ i. d.), CEC/MS could have the potential to greatly improve the poor concentration detection sensitivity of CE while at the same time retaining the high absolute sensitivity of CE.

The ability to measure m/z with high accuracy using time-of-flight (TOF) analyzers has improved significantly over the last few years. Hybrid mass spectrometers combining TOF mass analyzer with magnetic sector, quadrupole mass filter, and quadrupole ion trap to perform collision induced dissociation (CID). A benefit of TOF analyzer is a fast scan rate, good resolution, high mass accuracy and high sensitivity in full-scan mode. In combination with a quadrupole mass filter these features have been useful to derive partial and full sequence of peptides. CE could greatly benefit from the faster scan rate of TOF analyzer because of the narrow peak widths produced in CE.

The role of accurate mass measurement ($\pm 10\,ppm$) in protein identification strategies employing MS or MS/MS and database searching was described by Burlingame and co-workers [91]. We have investigated the impact of improved mass resolution and accuracy on the results obtained using the SEQUEST algorithm for searching database using CID spectra for

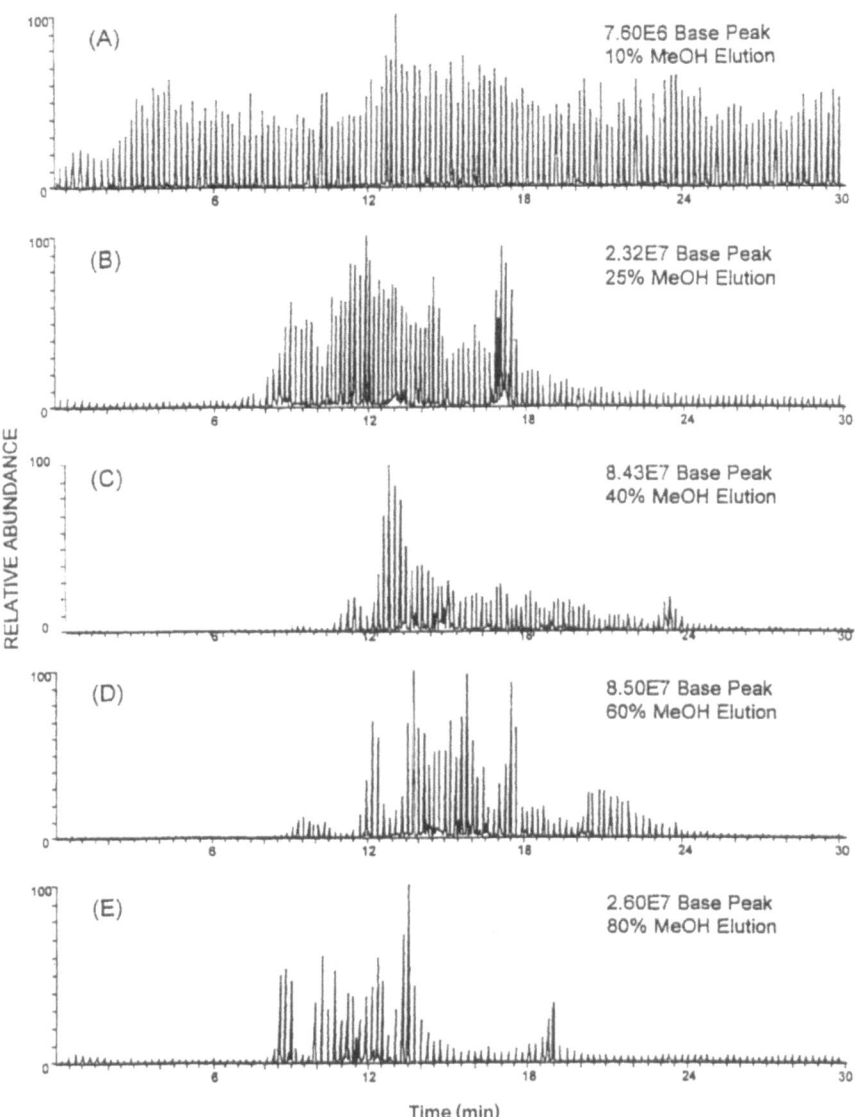

Figure 10. Micro-SPE/multistep elution/CE/MS/MS traces for the tryptic digest of yeast ribosomal complex. Elutions were performed by injecting different percentages of methanol (in 0.5% acetic acid) at $-25\,kV$ for 45 s, respectively. CE was run at $-20\,kV$ after each elution step and then reduced to $-8\,kV$ after methanol peak appeared. (**A**) 10, (**B**) 25, (**C**) 40, (**D**) 60, and (**E**) 80% methanol. (Reprinted with permission from *Anal. Chem.* **1999**, *71*, 2276.)

CE/ESI-in-source CID-TOF MS experiments [92]. Table II shows an obvious benefit of this mass measurement accuracy as a sequence delimiter in database searching. Searches using mass tolerances of 3.0, 1.0, 0.1 and 0.01 of the precursor ion m/z values were performed. Fragment ion mass tolerance was kept at 1.0 amu. No proteolytic cleavage specificity was used in the search. The number of sequences matching the molecular weight of the peptide fell from ~4.5 million for a 3 amu mass tolerance to ~23,700. By reducing the number of sequences evaluated, search times fell from 1 minute and 32 seconds to 38 seconds, an improvement of 58%. In a search of a small database (10,169 proteins), search times fell from 7

to 2 seconds, a 71% improvement. Scoring values produced by SEQUEST are impacted favorably leading to greater differences between the first and second ranked sequences for the correlation values.

Protein identification by CE/ESI-TOF MS using peptide mapping was reported [93]. Figeys et al reported the coupling of microfabricated device with an ESI-Q-TOF mass spectrometer for protein identification based on enhanced resolution MS and tandem MS data [94].

10.5 Conclusions

The high separation efficiency and sensitivity of CE when combined with the

Table I. Comparison of single-step and multistep elution micro-SPE/CE/MS/MS for protein identification of *S. cerevisiae* ribosomal proteins after digestion of the complex with trypsin[a]. (Reprinted with permission from *Anal. Chem.* **1999**, *71*, 2278.)

	Run No.	Unique Proteins Identified, no. (%)	Unique Peptides Scored	Proteins Identified with > 2 peptides	Proteins Identified with single peptide	Total No. of Peptides Scored
Single step	1[b]	38 (50.6%)	53	10	11	60
	2	39 (52%)	57	13	7	70
	3	31 (41.3%)	40	8	8	57
	4	36 (48%)	48	7	12	57
Multi-step	1[b,c]	55 (73%)	90	26	11	199
	2[d]	66 (88%)	136	38	11	476
	3[d]	63 (84%)	109	29	13	348
	4[e]	70 (93%)	122	34	13	380
	5[f]	53 (70.6%)	87	20	9	228
	6[g]	53 (70.6%)	92	25	8	182

[a] Unless otherwise stated, CE voltage was kept at −20 kV at first and reduced to −8 kV after methanol peak came out. [b] Using Microchrom reversed-phase packing material. [c] Eight steps (10, 20, 30, 40, 50, 60, 70, 80% MeOH). [d] 11 steps (10, 20, 25, 30, 35, 40, 45, 50, 60, 70, 80% MeOH). [e] Ten steps (2, 5, 10, 20, 30, 40, 50, 60, 70, 80% MeOH). [f] Seven steps (20, 30, 40, 50, 60, 70, 80% MeOH); CE voltage at −21 kV at first and reduced to −18 kV after MeOH peak came out. [g] Eight steps (same as a); constant CE voltage at −25 kV.

Table II. The impact of mass measurement accuracy on peptide database search.

Peptide Mass Tolerance		0.001	0.01	0.1	1.0	3.0
Non-redundant database (291,937)	Time Used	0:39	0:38	0:42	0:56	1:32
	# Matched Peptides	3906	23710	323609	1520339	4574024
	Time Used	0:02	–	–	–	0:07

[a] The search was performed on a DEC Alpha 500 MHz computer.

structural information provided by mass spectrometry, has become a powerful tool for protein identifications. Further development in this field will require improvement of concentration limits as well as the resolving power using multidimensional separation strategies. In this respect, the coupling of LC with CEC could potentially be more powerful than LC/CE. Microfabricated devices integrating sheathless nanospray source and a multidimensional separation mechanism will be another obvious avenue to more wide spread use of CE/MS/MS for sensitive and high resolution peptide separations in complex mixtures.

10.6 References

[1] Hjertén, S. *Chromatogr. Rev.* **1967**, *9*, 122.

[2] Mikkers, F.E.P.; Everaerts, F.M.; Verheggen, T.P.E.M. *J. Chromatogr.* **1979**, *169*, 11.

[3] Jorgenson, J.W.; Lukacs, K.D. *Anal. Chem.* **1981**, *53*, 1298.

[4] Messana, I.; Rossetti, D.V.; Cassiano, L.; Misiti, F.; Giardina, B.; Castagnola, M. *J. Chromatogr. B* **1997**, *699*, 149.

[5] Wehr, T.; Rodriguez-Diaz, R.; Zhu, M. *Capillary Electrophoresis of Proteins*, Marcel Dekker, New York, **1999**.

[6] Caprioli, R.M.; Moore, W.T., Martin, M.; DaGue, B.B. *J. Chromatogr.* **1989**, *480*, 247.

[7] Reinhoud, N.J.; Niessen, W.M.A.; Tjaden, U.R. *Rapid Commun. Mass Spectrom.* **1989**, *3*, 348.

[8] deWit, J.S.M.; Deterding, L.J.; Moseley, M.A.; Tomer, K.B.; Jorgenson, J.W. *Rapid Commun. Mass Spectrom.* **1988**, *2*, 100.

[9] Moseley, M.A.; Deterding, L.J.; Tomer, K.B.; Jorgenson, J.W. *J. Chromatogr.* **1989**, *480*, 197.

[10] Olivares, J.A.; Ngugen, N.T.; Yonker, C.R.; Smith, R.D. *Anal. Chem.* **1987**, *59*, 1230.

[11] Smith, R.D.; Olivares, J.A.; Ngugen, N.T.; Udseth, H.R. *Anal. Chem.* **1988**, *60*, 436.

[12] Ding, J.; Vouros, P. *Anal. Chem.* **1999**, *71*, 379A.

[13] Wilm, M.S.; Mann, M. *Int. J. Mass Spectrom. Ion Processes* **1994**, *136*, 2.

[14] Wahl, J.H.; Goodlett, D.R.; Udseth, H.R.; Smith, R.D. *Anal. Chem.* **1992**, *64*, 3194.

[15] Wahl, J.H.; Goodlett, D.R.; Udseth, H.R.; Smith, R.D. *Electrophoresis* **1993**, *14*, 448.

[16] Yates, J.R. III; McCormack, A.L.; Eng, J. *Anal. Chem.* **1996**, *68*, 534A.

[17] Dongré, A.R.; Eng, J.; Yates, J.R. III *Trends in Biotechnology* **1997**, *15*, 418.

[18] Yates, J.R. III *J. Mass Spectrom.* **1998**, *33*, 1.

[19] Eng, J.K.; McCormack, A.L.; Yates, J.R., III *J. Am. Soc. Mass Spectrom.* **1994**, *5*, 976.

[20] Yates, J.R. III; Eng, J.K. Use of mass spectrometry fragmentation patterns of peptides to identify amino acid sequences in database, U.S. Patent, 5, 538, 897, **1996**.

[21] Mann, M.; Wilm, M. *Anal. Chem.* **1994**, *66*, 4390.

[22] Clauser, K.R.; Hall, S.C.; Smith, D.M.; Webb, J.W.; Andrews, L.E.; Tran, H.M.; Epstein, L.B.; Burlingame, A.L. *Proc. Natl. Acad. Sci. U.S.A.* **1995**, *92*, 5072.

[23] Tong, W.; Link, A.; Eng, J.K.; Yates, J.R. III *Anal. Chem.* **1999**, *71*, 2270.

[24] Foret, F.; Thompson, T.J.; Vorous, P.; Karger, B.L.; Gebauer, P.; Bocek, P. *Anal. Chem.* **1994**, *66*, 4450.

[25] Johansson, I.M.; Huang, E.C.; Henion, J.D.; Zweigenbaum, J. *J. Chromatogr.* **1991**, *554*, 311.

[26] Fang, L.; Zhang, R.; Zare, R.N. *Proc. 41st ASMS Conf. Mass Spectrometry and Allied Topics*, 755a, **1993**.

[27] Kriger, M.S.; Cook, K.D.; Ramsey, R.S. *Anal. Chem.* **1995**, *67*, 385.

[28] Ramsey, R.S.; McLuckey, S.A. *J. Microcol. Sep.* **1995**, *7*, 461.

[29] Figeys, D.; Oostveen, I.; Ducret, A.; Aebersold, R. *Anal. Chem.* **1996**, *68*, 1822.

[30] Seuers, J.L.; Harms, A.C.; Smith, R.D. *Rapid Commun. Mass Spectrom.* **1996**, *10*, 1175.

[31] Bateman, K.P.; White, R.L.; Thibault, P. *Rapid Commun. Mass Spectrom.* **1997**, *11*, 307.

[32] Figeys, D.; Ducret, A.; Aebersold, R. *J. Chromatogr.* **1997**, *763*, 295.

[33] Gatlin, C.L.; Kleeman, G.R.; Hays, L.G.; Link, A.J.; Yates, J.R. III *Anal. Biochem.* **1998**, *263*, 93.

[34] Cao, P.; Moini, M. *J. Am. Soc. Mass Spectrom.* **1997**, *8*, 561.

[35] Tong, W.; Yates, J.R. III *Unpublished results.*

[36] Mazereeuw, M.; Hofte, A.J.P.; Tjaden, U.R.; Greef, J. *Rapid Commun. Mass Spectrom.* **1997**, *11*, 981.

[37] Manz, A.; Becker, H. Ed. *Microsystem Technologies in Chemistry and Life Science*, Springer-Verlag, Berlin, **1998**.

[38] Xue, Q.; Foret, F.; Dunayevskiy, Y.M.; Karger, B.L. *Rapid Commun. Mass Spectrom.* **1997**, *11*, 1253.

[39] Xue, Q.; Foret, F.; Dunayevskiy, Y.M.; Karger, B.L. *Anal. Chem.* **1997**, *69*, 426.

[40] Ramsey, R.S.; Ramsey, J.M. *Anal. Chem.* **1997**, *69*, 1174.

[41] Figeys, D.; Ning, Y.; Aebersold, R. *Anal. Chem.* **1997**, *69*, 3153.

[42] Bings, N.H.; Wang, C.; Skinner, C.D. Colyer, C.L.; Thibault, P.; Harrison, D.J. *Anal. Chem.* **1999**, *71*, 3292.

[43] Zhang, B.; Liu, H.; Karger, B.L.; Foret, F. *Anal. Chem.* **1999**, *71*, 3258.

[44] Li, J.; Thibault, P.; Bing, N.H.; Skinner, C.D.; Wang, C.; Colyer, C.; Harrison, *J. Anal. Chem.* **1999**, *71*, 3036.

[45] Figeys, D.; Gigy, S.P.; McKinnon, G.; Aebersold, R. *Anal. Chem.* **1998**, *70*, 3728.

[46] Figeys, D.; Aebersold, R. *Anal. Chem.* **1998**, *70*, 3721.

[47] Lazer, I.M.; Ramsey, R.S.; Sundberg, S.; Ramsey, J.M. *Anal. Chem.* **1999**, *71*, 3627.

[48] Valaskovic, G.A.; Kelleher, N.L.; Little, D.P.; Aaserud, D.J.; McLafferty, F.W. *Anal. Chem.* **1995**, *67*, 3802.

[49] Hannis, J.C.; Muddiman, D.C. *Rapid Commun. Mass Spectrom.* **1998**, *12*, 443.

[50] Foret, F.; Szoko, E.; Karger, B.L. *Electrophoresis* **1993**, *14*, 417.

[51] Thompson, T.; Foret, F.; Vouros, P.; Karger, B.L. *Anal. Chem.* **1993**, *65*, 900.

[52] Tang, Q.; Harrata, A.K.; Lee, C.S. *Anal. Chem.* **1997**, *69*, 3177.

[53] Clarke, N.J.; Tomlinson, A.J.; Naylor, S. *J. Am. Soc. Mass Spectrom.* **1997**, *8*, 743.

[54] Figeys, D.; Ducret, A.; Yates, J.R. III; Aebersold, R. *Nature Biotechnol.* **1996**, *14*, 1579.

[55] Debets, D.S.; Mazereeuw, M.; Voogt, W.H.; Van Iperen, D.J.; Lingeman, H.; Hupe, K.P.; Brinkman, U.A.T. *J. Chromatogr.* **1992**, *608*, 151.

[56] Tomlinson, A.J.; Guzman, N.A.; Naylor, S. *J. Cap. Electrophor.* **1995**, *2*, 247.

[57] Tomlinson, A.J.; Benson, L.M.; Braddock, W.D.; Oda, R.P.; Naylor, S. *J. High Resolut. Chromatogr.* **1994**, *17*, 729.

[58] Tomlinson, A.J.; Naylor, S. *J. High Resolut. Chromatogr.* **1995**, *18*, 384.

[59] Tomlinson, A.J.; Naylor, S. *J. Liq. Chromatogr.* **1995**, *18*, 3591.

[60] Tomlinson, A.J.; Benson, L.M.; Jameson, S.; Naylor, S. *Electrophoresis* **1996**, *17*, 1801.

[61] Tomlinson, A.J.; Benson, L.M.; Jameson, S.; Johnson, D.H.; Naylor, S. *J. Am. Soc. Mass Spectrom.* **1997**, *8*, 15.

[62] Kurian, E.; Prendergast, F.G.; Tomlinson, A.J.; Holmes, M.W.; Naylor, S. *J. Am. Soc. Mass Spectrom.* **1997**, *8*, 8.

[63] Goodlett, D.R.; Wahl, J.H.; Udseth, H.R.; Smith, R.D. *J. Microcol. Sep.* **1993**, *5*, 57.

[64] Moseley, M.A.; Deterding, L.J.; Tomer, K.B.; Jorgenson, J.W. *Anal. Chem.* **1991**, *63*, 109.

[65] Bruin, G.J.; Huisden, R.; Kraak, J.C.; Poppe, H. *J. Chromatogr.* **1989**, *480*, 339.

[66] Li., M.X.; Liu, L.; Wu, J.T.; Lubman, D.M. *Anal. Chem.* **1997**, *69*, 2451.

[67] Thibault, P.; Paris, C.; Pleasance, S. *Rapid Commun. Mass Spectrom.* **1991**, *5*, 484.

[68] Kelly, J.F.; Locke, S.J.; Ramaley, L.R.; Thibault, P. *J. Chromatogr.* **1996**, *720*, 409.

[69] Bateman, K.P.; Thibault, P.; Douglas, D.J.; White, R.L. *J. Chromatogr.* **1995**, *712*, 253.

[70] Kelly, J.F.; Ramaley, L.; Thibault, P. *Anal. Chem.* **1997**, *69*, 51.

[71] Figeys, D.; Carthals, G.L.; Gallis, B.; Goodlett, D.R.; Ducret, A.; Corson, M.A.; Aebersold, R. *Anal. Chem.* **1999**, *71*, 2279.

[72] McCormack, A.L.; Schieltz, D.M.; Goode, B.; Yang, S.; Barnes, G.; Drubin, D.; Yates, J.R., III *Anal. Chem.* **1997**, *69*, 767.

[73] Link, A.J.; Eng, J.; Schieltz, D.M.; Carmack, E.; Mize, G.J.; Morris, D.R.; Garvik, B.M.; Yates, J.R., III *Nature Biotechnol.* **1999**, *17*, 676.

[74] Giddings, J.C. *Anal. Chem.* **1984**, *56*, 1258A.

[75] Jeffery, D.J.; Hooker, T.F.; Jorgenson, J.W. in *Handbook of Capillary Electrophoresis*, 2nd Ed., Landers, J.P. Ed., CRC Press, Boca Raton, **1997**, p.765.

[76] Bushey, M.M.; Jorgenson, J.W. *Anal. Chem.* **1990**, *62*, 978.

[77] Bushey, M.M.; Jorgenson, J.W. *J. Microcol. Sep.* **1990**, *2*, 293.

[78] Larmann, J.P. Jr.; Lemmo, A.V.; Moore, A.W. Jr.; Jorgenson, J.W. *Electrophoresis* **1993**, *14*, 439.

[79] Lemmo, A.V.; Jorgenson, J.W. *Anal. Chem.* **1993**, *65*, 1576.

[80] Moore, A.W., Jr.; Jorgenson, J.W. *Anal. Chem.* **1995**, *67*, 3448.

[81] Lemmo, A.V.; Jorgenson, J.W. *J. Chromatogr.* **1993**, *633*, 213.

[82] Moore, A.W., Jr.; Jorgenson, J.W. *Anal. Chem.* **1995**, *67*, 3456.

[83] Yamamoto, H.; Manabe, T.; Okuyamam T. *J. Chromatogr.* **1989**, *480*, 277.

[84] Yamamoto, H.; Manabe, T.; Okuyamam T. *J. Chromatogr.* **1989**, *480*, 277.

[85] Lewis, K.C.; Opiteck, G.J.; Jorgenson, J.W.; Sheeley, D.M. *J. Am. Soc. Mass Spectrom.* **1997**, *8*, 495.

[86] Guzman, N.A. *LC-GC* **1999**, *17*, 16.

[87] Colón, L.A.; Guo, Y.; Fermier, A. *Anal. Chem.* **1997**, *69*, 461A.

[88] Huang, P.; Li, M.X.; Qian, M.G.; Lubman, D.M. *Anal. Chem.* **1997**, *69*, 320.

[89] Wu, J.T.; Huang, P.; Li, M.X.; Lubman, D.M. *Anal. Chem.* **1997**, *69*, 2908.

[90] Huang, X.; Gusev, I.; Horváth, C. 12th International Symposium on High Performance Capillary Electrophoresis & Related Microscale Techniques, Palm Spring, CA, **1999**, p. 104.

[91] Clauser, K.P.; Baker, P.; Burlingame, A.L. *Anal. Chem.* **1999**, *71*, 2871.

[92] Eng, J.K.; Tong, W.; Yates, J.R. III, Unpublished results.

[93] Cao, P.; Moini, M. *Rapid Commun. Mass Spectrom.* **1998**, *12*, 864.

[94] Figeys, D.; Locke, C.; Taylor, L.; Aebersold, R. *Rapid Commun. Mass Spectrom.* **1998**, *12*, 1435.

Chromatographia Supplement Vol. 53, 2001, Subject Index

Migration Time, **53**, S-36, S-71
–, Precision, **53**, S-36, S-71
Monoclonal Antibody, **53**, S-13, S-14, S-50
–, CE-SDS, **53**, S-60, S-78
–, cIEF, **53**, S-50
–, Synagis®, **53**, S-66
–, Analysis, **53**, S-50
Monosaccharides, **53**, S-85
–, Determination, **53**, S-88

Offord Model, **53**, S-9, S-31

Peak Area, **53**, S-36
–, Reproducibility (Precision), **53**, S-36
Peptides, **53**, S-29
–, Buffer selection for CZE, **53**, S-18
–, Peptide maps, **53**, S-11
–, Protegrin analogs, **53**, S-28
p*I* determination, **53**, S-88
PNGase F, **53**, S-40
Preparative IEF, **53**, S-60
Protein
–, Deoxyribonuclease, **53**, S-39
–, Enbrel®, **53**, S-60
–, Erythropoietin, **53**, S-12, S-13, S-57

–, Granulocyte colony stimulating factor, **53**, S-13
–, Granulocyte macrophage colony stimulating factor, **53**, S-11
–, Growth hormone, **53**, S-11
–, Hemoglobin, **53**, S-54
–, Human serum albumin, **53**, S-34
–, Human bone morphogenic protein-2, **53**, S-11
–, Infergen, **53**, S-35
–, Insulin, **53**, S-13
–, Interleukin, **53**, S-10
–, rhuF(ab')$_2$, **53**, S-77
–, rhuMab, **53**, S-13, S-14, S-50
–, rhuTNFR:Fc, **53**, S-60
–, Recombinant proteins, **53**, S-11
–, Synagis®, **53**, S-66
–, Tissue plasminogen activator, **53**, S-11
Protein therapeutics, **53**, S-75
Purity, purification, **53**, S-13

Quality Control, **53**, S-15, S-66
Quantification, **53**, S-13, S-35

Recombinant proteins, **53**, S-11
Reverse-Phase HPLC, **53**, S-27

–, Separation of protegrin analogs, **53**, S-27

SDS-PAGE, **53**, S-67
Stability Studies, **53**, S-14

Validation, **53**, S-68
–, Parameters, **53**, S-68
–, – Accuracy, **53**, S-53, S-70
–, – Intermediate precision, **53**, S-36, S-70
–, – Linearity, **53**, S-53, S-69
–, – Range, **53**, S-69
–, – Repeatability, **53**, S-70
–, – Reproducibility, **53**, S-36, S-71
–, – Robustness, **53**, S-53, S-71
–, – Sensitivity, **53**, S-71
–, – Specificity, **53**, S-53, S-68
–, – System suitability, **53**, S-72
–, Pilot studies, **53**, S-68
–, Protocol, **53**, S-68
–, Report, **53**, S-72
–, Tips, **53**, S-72

UV Detection, **53**, S-16, S-86

List of Abbreviations

BGE	Background electrolyte
BME	β-Mercaptoethanol
BSA	Bovine serum albumin
CCD	Charge coupled device
CE	Capillary electrophoresis
CE-SDS	Capillary electrophoresis-Sodium dodecylsulfate
cIEF	Capillary isoelectric focusing
CGE	Capillary gel electrophoresis
CHAPS	3-[(3-Cholamidopropyl)dimethylammonio]-1-propanesulfonate
CZE	Capillary zone electrophoresis
DAB	1,4-Diaminobutane
DNase	Deoxyribonuclease
DTT	Dithiothreitol
EOF	Electroendoosmotic flow
ELISA	Enzyme-linked immunosorbent assay
EPO	Erythropoietin
ESI	Electrospray ionization
h	Human
hGH	Human growth hormone
HPLC	High performance liquid chromatography
HPMC	Hydroxylpropylmethyl cellulose
HSA	Human serum albumin
ICH	International conference on harmonization
IEF	Isolectric focusing
LC/HC	Light chain/Heavy chain (of immunoglobulins)
LC-MS	Liquid chromatography-Mass spectrometry
LIF	Laser-induced fluorescence
LOD	Limit of detection
MAb	Monoclonal antibody
MALDI-TOF	Matrix assisted laser desorption ionization-time of flight
MECC	Micellar electrokinetic capillary chromatography (also known as MEKC)
MEKC	Micellar electrokinetic chromatography
MS	Mass Spectrometry
OOS	Out of specification
pH	$-\log[H^+]$
pI	Isoelectric point
PNGase F	Peptide-N-glycanase F
PVA	Polyvinylalcohol
r	Recombinant
RPHPLC	Reverse-phase HPLC
RSD	Relative standard deviation
SDS-PAGE	SDS-polyacrylamide gel electrophoresis
SEC	Size-exclusion chromatography
SPE	Solid phase extraction
TEMED	N,N,N',N',Tetramethylenediamine
t-PA	Tissue plasminogen activator
UV	Ultraviolet
2D	Two-dimensional